189

Advances in Polymer Science

Advances in Polymer Science

Recently Published and Forthcoming Volumes

Surface-Initiated Polymerization II
Volume Editor: Jordan, R.
Vol. 198, 2006

Surface-Initiated Polymerization I
Volume Editor: Jordan, R.
Vol. 197, 2006

Conformation-Dependent Design of Sequences in Copolymers II
Volume Editor: Khokhlov, A. R.
Vol. 196, 2006

Conformation-Dependent Design of Sequences in Copolymers I
Volume Editor: Khokhlov, A. R.
Vol. 195, 2006

Enzyme-Catalyzed Synthesis of Polymers
Volume Editors: Kobayashi, S., Ritter, H., Kaplan, D.
Vol. 194, 2006

Polymer Therapeutics II
Polymers as Drugs, Conjugates and Gene Delivery Systems
Volume Editors: Satchi-Fainaro, R., Duncan, R.
Vol. 193, 2006

Polymer Therapeutics I
Polymers as Drugs, Conjugates and Gene Delivery Systems
Volume Editors: Satchi-Fainaro, R., Duncan, R.
Vol. 192, 2006

Interphases and Mesophases in Polymer Crystallization III
Volume Editor: Allegra, G.
Vol. 191, 2005

Block Copolymers II
Volume Editor: Abetz, V.
Vol. 190, 2005

Block Copolymers I
Volume Editor: Abetz, V.
Vol. 189, 2005

Intrinsic Molecular Mobility and Toughness of Polymers II
Volume Editor: Kausch, H.-H.
Vol. 188, 2005

Intrinsic Molecular Mobility and Toughness of Polymers I
Volume Editor: Kausch, H.-H.
Vol. 187, 2005

Polysaccharides I
Structure, Characterization and Use
Volume Editor: Heinze, T.
Vol. 186, 2005

Advanced Computer Simulation Approaches for Soft Matter Sciences II
Volume Editors: Holm, C., Kremer, K.
Vol. 185, 2005

Crosslinking in Materials Science
Vol. 184, 2005

Phase Behavior of Polymer Blends
Volume Editor: Freed, K.
Vol. 183, 2005

Polymer Analysis/Polymer Theory
Vol. 182, 2005

Interphases and Mesophases in Polymer Crystallization II
Volume Editor: Allegra, G.
Vol. 181, 2005

Interphases and Mesophases in Polymer Crystallization I
Volume Editor: Allegra, G.
Vol. 180, 2005

Block Copolymers I

Volume Editor: Volker Abetz

With contributions by
V. Abetz · N. Hadjichristidis · H. Iatrou
M. Pitsikalis · P. F. W. Simon

Springer

The series *Advances in Polymer Science* presents critical reviews of the present and future trends in polymer and biopolymer science including chemistry, physical chemistry, physics and material science. It is adressed to all scientists at universities and in industry who wish to keep abreast of advances in the topics covered.
As a rule, contributions are specially commissioned. The editors and publishers will, however, always be pleased to receive suggestions and supplementary information. Papers are accepted for *Advances in Polymer Science* in English.
In references *Advances in Polymer Science* is abbreviated *Adv Polym Sci* and is cited as a journal.

Springer WWW home page: http://www.springeronline.com
Visit the APS content at http://www.springerlink.com/

ISBN 978-3-642-06589-7 e-ISBN 978-3-540-31602-2

ISSN 0065-3195

DOI 10.1007/b137234

Springer is a part of Springer Science+Business Media

springeronline.com

Softcover reprint of the hardcover 1st edition 2005

Cover design: *Design & Production* GmbH, Heidelberg

Printed on acid-free paper 02/3141 YL – 5 4 3 2 1 0

Volume Editor

Prof. Dr. Volker Abetz

GKSS – Forschungszentrum Geesthacht GmbH
Institut für Polymerforschung
Max-Planck-Straße
21502 Geesthacht, Germany
Volker.Abetz@gkss.de

Editorial Board

Advances in Polymer Science Also Available Electronically

For all customers who have a standing order to Advances in Polymer Science, we offer the electronic version via SpringerLink free of charge. Please contact your librarian who can receive a password or free access to the full articles by registering at:

springerlink.com

If you do not have a subscription, you can still view the tables of contents of the volumes and the abstract of each article by going to the SpringerLink Homepage, clicking on "Browse by Online Libraries", then "Chemical Sciences", and finally choose Advances in Polymer Science.

You will find information about the

- Editorial Board
- Aims and Scope
- Instructions for Authors
- Sample Contribution

at springeronline.com using the search function.

Preface

Self-assembly of matter into ordered structures is an important issue in biology and materials science. For example, the intramolecular and intermolecular self-assembly of proteins is essential for their functions as enzymes or carrier systems. In materials science, the organization of matter is important for mechanical, optical, electrical, and other properties not only in bulk or thin film systems, but also in solution. Self-assembly is a property of some materials which can be used for the controlled generation of regularly structured materials. This self-assembly can occur on different length scales. Atoms, small molecules or repeating units of some polymers can aggregate into crystals, which have typical periodicities on the sub-nanometer range. On a much larger length scale another class of materials becomes interesting for its capability to self-assemble into ordered structures: block copolymers. They are a fascinating class of condensed soft matter. By linking different, mostly immiscible chains together chemically, they self-assemble often into crystal-like structures (supercrystals, microphase morphologies). The periodic length scale in microphase separated block copolymers is typically in the range between 10 and several 100 nm, i.e. 100 to 1000 times larger than the periodic length of atomic crystals.

Block copolymers have already attracted significant scientific and economic interest during the last few decades. A prerequisite for the formation of rather periodic microphase morphologies is a fairly narrow polydispersity both of composition and molecular weight. After the discovery of living anionic polymerization in the middle of the last century, the controlled synthesis of block copolymers became possible for the first time. Since then, other synthetic schemes with a large degree of control were developed, and this trend is still continuing. For this reason more and more monomers become eligible for incorporation into block copolymers, thus adding more possible functions into microphase-structured materials.

Although many books and reviews have been written about block copolymers, the ongoing research motivated us to give an overview of the results during the last years in various aspects of the field.

The topic is split into two volumes and is organized as follows. In the first chapter the progress in different synthetic routes to controlled block copolymers of various architectures will be presented, the second chapter tries to give an overview of the phase behavior of block copolymers in the bulk state and in concentrated solution. The interplay between crystallization on a segmental scale and the microphase separated structure of block copolymers with crystallizable blocks will be discussed extensively in the third chapter, since this aspect has not been addressed in such detail before in a review. After having treated bulk or bulk-like (highly concentrated solutions) states, the fourth chapter presents the structure formation of block copolymers in more dilute solutions, where various micellar superstructures can be found. The last two chapters deal with applications of block copolymer structures as precursors for the formation of mesoscale porosity (fifth chapter), and as precursors for the formation of controlled patterns on surfaces enabling new ways to lithography on a length scale, which is not accessible by other lithographic techniques thus far (sixth, last chapter).

We are aware that these two volumes do not cover all aspects of the research done on block copolymers, but nevertheless it is our hope that it will find interested readers and be a basis and stimulation for future further research on these materials.

Geesthacht, October 2005 *Volker Abetz*

Contents

Contents of Volume 190

Block Copolymers II

Volume Editor: Volker Abetz
ISBN: 3-540-26902-9

Adv Polym Sci (2005) 189: 1–124
DOI 10.1007/12_005

Published online: 20 October 2005

Synthesis of Block Copolymers

Nikos Hadjichristidis (✉) · Marinos Pitsikalis · Hermis Iatrou

Department of Chemistry, University of Athens, Panepistimiopolis Zografou, 15771 Athens, Greece
hadjichristidis@chem.uoa.gr

Abstract This review highlights recent (2000-2004) advances and developments regarding the synthesis of block copolymers with both linear [AB diblocks, ABA and ABC triblocks, ABCD tetrablocks, (AB)n multiblocks etc.] and non-linear structures (star-block, graft, miktoarm star, H-shaped, dendrimer-like and cyclic copolymers). Attention is given only to those synthetic methodologies which lead to well-defined and well-characterized macromolecules.

Abbreviations

AFM	atomic force microscopy
AIBN	α,α'-azobisisobutyronitrile
AMBA	sodium 3-acrylamido-3-methylbutanoate
AMPS	sodium 2-acrylamido-2-methylpropane sulfonate
ATRP	atom transfer radical polymerization
BDSIMP	bis(*tert*-butyldimethylsilyloxymethylphenyl)
BMBP	2,2-bis(methylene α-bromopropionate)propionyl
bpy	4,4′-dialkyl substituted bipyridine
BzMA	benzyl methacrylate
CbzNB	5-(*N*-carbazoyl methylene)-2-norbornene
CDMSS	4-(chlorodimethylsilyl)styrene
CMP	chloromethyl phenyl-
CMS	chloromethylstyrene
CTA	chain transfer agent
D_3	hexamethylcyclotrisiloxane
DABCO	1,4-diazabicyclo[2,2,2]octane
DCC	dicyclohexyl carbodiimide
DDPE	double diphenyl ethylenes
DEPN	*N*-t-butyl-*N*-(1-diethylphosphono-2,2-dimethylpropyl)nitroxide
DLLA	D,L-lactide
DMAP	4-(dimethylamino) pyridine
DME	dimethoxyethane
DMF	dimethylformamide
DMSO	dimethyl sulfoxide
DMVBAC	*N,N′*-dimethylvinylbenzylamine
DPE	1,1-diphenylethylene
DPMK	diphenylmethylpotassium
DPQ	2,3-dichloro-5,6-dicyano-1,4-benzoquinone
DSC	differential scanning calorimetry
DTPA	dithiobis(propionic acid)
DVB	divinylbenzene
DVC	divinyl compound
FMA	2-(*N*-methylperfluorobutane sulfonamide)ethyl methacrylate
GAMA	2-gluconamido ethyl methacrylate
GMA	glycerol monomethacrylate
GTP	group transfer polymerization
HBBIB	4-hydroxy-butyl-2-bromoisobutyrate
HEBB	β-hydroxyethyl α-bromobutyrate
HES	hexaepoxysqualene
HMTETA	1,1,4,7,10,10-hexamethyl triethylenetetramine
IPP	isopropenylpyridine

Is	isoprene
LALLS	low angle laser light scattering
LAMA	2-lactobionamido ethyl methacrylate
LFRP	living free radical polymerization
MAHE	*trans, trans*-1-methacryloyloxy-2,4-hexadiene
MALLS	multi angle laser light scattering
MAO	methylaluminoxane
MBA	3-methacryloyloxy-1,1′-biadamantane
Me_6-TREN	[(2-dimethylamino)ethyl]amine
MHI	multifunctional initiator
MLA	multifunctional linking agent
M_n	number-average molecular weight
MPEO	poly(ethylene oxide) monomethyl ether
MTS	1-methoxy-1-trimethylsiloxy-2-methyl-1-propane
M_w	weight-average molecular weight
M_w/M_n	molecular weight distribution
NIPA	*N*-isopropyl acrylamide
NMP	nitroxide mediated polymerization
NMR	nuclear magnetic resonance
OEGMA	oligo(ethylene glycol) monomethylether methacrylate
P2MP	poly(2-methyl-1,4-pentadiene)
P2VP	poly(2-vinylpyridine)
P4VP	poly(4-vinylpyridine)
P(7CC)	poly(1,3-dioxepan-2-one)
PAAM	polyacrylamide
PBBOS	poly{2,5-bis[(4-butylbenzoyl)oxy]styrene}
PBd	polybutadiene
PCHD	poly(1,3-cyclohexadiene)
PCL	poly(*ε*-caprolactone)
PCMS	poly(*p*-chloromethyl styrene)
PCP	polycyclopentene
PD	polydiene
PDEAEMA	poly[2-(diethylamino)ethyl methacrylate]
PDIPAEMA	poly[2-(diisopropylamino)ethyl methacrylate]
PDMA	Poly(decyl methacrylate)
PDMAA	poly(*N,N*-dimethylacrylamide)
PDMAEMA	poly[2-(dimethylamino)ethyl methacrylate]
PDMS	poly(dimethylsiloxane)
PDOP	poly(1,3-dioxepane)
PE	polyethylene
PEG	poly(ethylene glycol)
PEMA	poly(ethyl methacrylate)
PEN	poly(ethylidenenorbornene)
PEO	poly(ethylene oxide)
PEP	poly(ethylene-*alt*-propylene)
PF	polyfluorene
PFS	poly(ferrocenyl dimethyl silane)
PHEGMA	poly[hexa(ethylene glycol) methacrylate]
PHOS	poly(*p*-hydroxy styrene)
PHPMA	poly[*N*-(2-hydroxypropyl) methacrylamide]

PHS	poly(p-hydroxystyrene)
PI	polyisoprene
PIB	polyisobutylene
PIBVE	poly(isobutyl vinyl ether)
PIsOA	Poly(isooctyl acrylate)
PLLA	poly(L-lactide)
PMA	poly(methyl acrylate)
PMDETA	*N,N,N′,N′,N″*-pentamethyl diethylenetriamine
PMEMA	poly[2-(*N*-morpholino)ethyl methacrylate]
PMeOx	poly(2-methyl oxazoline)
PMMA	poly(methyl methacrylate)
PMOS	poly(*p*-methoxystyrene)
PMVE	poly(methyl vinyl ether)
PNBD	polynorbornadiene
P*n*BuA	Poly(*n*-butyl acrylate)
POHVE	poly(2-hydroxyethyl vinyl ether)
PPO	poly(propylene oxide)
PS	polystyrene
PSMA	poly(stearyl methacrylate)
PSPMA	poly(sulfopropyl methacrylate)
P*t*BuA	poly(*tert*-butyl acrylate)
P*t*BuMA	poly(*tert*-butyl methacrylate)
P*t*BuS	poly(*t*-butylstyrene)
PTHPMA	poly(tetrahydropyranyl methacrylate)
PTMSMA	poly(trimethyl silyl methacrylate)
PVAc	poly(vinyl acetate)
PVL	poly(δ-valerolactone)
PVP	poly(*N*-vinyl pyrrolidone)
PαMeS	poly(α-methyl styrene)
ROMP	ring opening metathesis polymerization
ROP	ring opening polymerization
s-BuLi	*sec*-butyllithium
SEC	size exclusion chromatography
SLS	static light scattering
SPP	3-[*N*-(3-methacrylamidopropyl)-*N,N*-dimethyl]ammoniopropane sulfonate
St	styrene
$Sn(Oct)_2$	stannous octoate
TBABB	tetra-*n*-butyl ammonium bibenzoate
*t*BuLi	*tert*-butyllithium
TEA	triethylamine
TEMPO	2,2,6,6-tetramethyl-1-piperidinyloxy stable radical
THF	tetrahydrofuran
TMEDA	*N,N,N′,N′*-tetramethylethylenediamine
UV	ultra violet spectroscopy

1
Introduction

Block copolymers are macromolecules composed of linear or nonlinear arrangements of chemically[1] different polymeric chains (blocks). In most cases the different blocks are incompatible, giving rise to a rich variety of well-defined self-assembled structures both in bulk and selective solvents. These self-assembled structures are the basis for applications ranging from thermoplastic elastomers to information storage, drug delivery, and photonic materials. As a result, there is a continuous investigation of the self-assembly processes as well as of the response of these materials to external stimuli. Therefore, it is not surprising that these materials play a central role in contemporary macromolecular science covering the full spectrum of polymer chemistry, polymer physics, and applications.

Several excellent books and review articles have been published covering this particular area of polymer science [1–3]. Nevertheless, this review will highlight recent (2000–2004) advances and developments regarding the synthesis of block copolymers with both linear (AB diblocks, ABA and ABC triblocks, ABCD tetrablocks, $(AB)_n$ multiblocks etc.) and non-linear structures (star-block, graft, miktoarm star, H-shaped, dendrimer-like, and cyclic copolymers). Attention will be given only to those synthetic methodologies which lead to well-defined and well-characterized macromolecules.

2
Synthesis of Linear Block Copolymers

2.1
General Synthetic Strategies

An indispensable requirement for the preparation of well-defined block copolymer structures is the utilization of a living, or at least a controlled chain-growth polymerization method, in connection with suitable purification methods for all reagents employed (monomers, solvents, linking agents, additives etc.) and techniques for excluding the introduction of any impurity in the polymerization system. Under such conditions undesired termination or transfer reactions are absent, or at least minimized allowing for the synthesis of chemically and molecularly homogeneous structures.

Two methods have been developed for the synthesis of AB diblock copolymers: (a) sequential addition of monomers; and (b) coupling of two appropriately end-functionalized chains. The first method is the most widely used

[1] Stereoblock copolymers with blocks of the same monomeric unit but different configurations are out of the scope of this review.

for the synthesis of block copolymers. An essential consideration for the successful employment of the technique is the order of monomer addition. The living chain from the polymerization of the first monomer must be able to efficiently initiate the polymerization of the second monomer. Another important requirement is that the conversion of the first monomer must be quantitative in order to achieve control over the molecular weights as well as chemical and structural homogeneity.

The synthesis of ABA triblock copolymers can be accomplished using one of the following methods: (a) three-step sequential addition of monomers; (b) two-step sequential addition of monomers followed by a coupling reaction with a suitable difunctional linking agent; and (c) use of a difunctional initiator and a two-step sequential addition of monomers. The purification requirements mentioned above are also valid for the successful use of the three-step sequential addition. This method has the advantage that asymmetric triblock copolymers ABA′, with the two A blocks having different molecular weights, can be prepared. The second method involves the synthesis of the living diblock AB, where the B block has only half the molecular weight compared to the desired one, followed by reaction with a difunctional linking agent. The linking reaction must be efficient and fast. However, the stoichiometry of the reaction is difficult to control. Consequently, a small excess of the living diblock copolymer is used to ensure complete linking. The excess diblock copolymer has to be removed by fractionation in a later step. This technique can be only used for the synthesis of symmetric ABA triblock copolymers. The third method is limited to the synthesis of symmetric triblock copolymers, but is versatile, since it involves only a two-step reaction without fractionation or other purification steps. The main limitation, however, of this method is the choice of the difunctional initiator, which must be able to initiate the polymerization of the desired monomer with the same rate from both directions.

Other more complex linear block co-, ter- and quarterpolymers, such as ABC, ABCD, ABABA can be prepared using the previously mentioned methods. An important tool in the synthesis of block copolymers involves the use of post-polymerization chemical modification reactions. These reactions must be performed under mild conditions to avoid chain scission, crosslinking, or degradation, but facile enough to give quantitative conversions. Hydrogenation, hydrolysis, hydrosilylation and quaternization reactions are among the most important post-polymerization reactions used for the preparation of block copolymers.

2.2
Synthesis of Block Copolymers by Anionic Polymerization

Since the discovery of living anionic polymerization and the first synthesis of diblock copolymers, this technique has emerged as the most reliable and versatile tool for the synthesis of model polymers having controlled architectures, microstructure and molecular weights, narrow molecular weight distributions, and chemical and compositional homogeneity [4, 5]. Under the appropriate experimental conditions anionic polymerization is associated with the absence of any spontaneous termination or chain transfer reaction, leading to the preparation of well-defined structures. Several initiators, monofunctional, difunctional, or multifunctional, along with different series of suitable linking agents having various functionalities are available for the synthesis of complex macromolecular architectures [6, 7]. An important limitation of anionic polymerization is the demanding experimental conditions required to achieve a living polymerization system [8] and its applicability to a rather narrow spectrum of monomers (styrenes, dienes, methacrylates, acrylates, ethylene oxide, vinyl pyridines). However, recent developments have allowed for the expansion of the utility of the method to a broad range of monomers such as methacrylates with bulky or functional ester groups, lactones, hexamethylcyclotrisiloxane, 1,3-cyclohexadiene, isocyanates etc.

ABA triblock copolymers, where A is poly(methyl methacrylate), PMMA, and B is poly(*t*-butyl acrylate), P*t*BuA, were prepared by anionic polymerization and sequential addition of monomers [9, 10]. The polymerization was conducted in tetrahydrofuran, THF, at – 78 °C using 1,1-diphenyl-3-methylpentyllithium as the initiator, prepared *in situ* by the reaction of *s*-BuLi and 1,1-diphenylethylene, DPE. LiCl was used as an additive to promote the living polymerization of the (meth)acrylates and lead to well-defined products having narrow molecular weight distributions. Selective acid-catalyzed transalcoholysis at 150 °C of the *t*-butyl ester groups by either *n*-butanol or isooctanol transformed the initial triblock copolymers PMMA-*b*-PtBuA-*b*-PMMA to PMMA-*b*-poly(*n*-butyl acrylate)-*b*-PMMA, PMMA-*b*-P*n*BuA-*b*-PMMA, and PMMA-*b*-poly(isooctyl acrylate)-*b*-PMMA, PMMA-*b*-PIsOA-*b*-PMMA triblocks, respectively. NMR analysis revealed that the transalcoholysis reaction was selective and almost quantitative (95–98%). With this procedure thermoplastic elastomers can be prepared with PMMA as the high Tg end-blocks and either P*n*BuA or PIsOA as the soft and low Tg middle block. These materials are expected to provide oxidative stability superior to the classical thermoplastic elastomers PS-*b*-PD-*b*-PS where PD is a polydiene. However, the greater miscibility of the methacrylate blocks with the acrylic ones compared to the PS and PD blocks leads to inferior mechanical properties.

Block copolymers comprised of PS and polymethacrylate blocks with aliphatic stearyl or decyl side groups were prepared by the sequential addition of monomers, as shown in Scheme 1. Styrene was polymerized in THF at – 78 °C using *s*-BuLi as the initiator [11, 12]. The nucleophilicity of the living polystyryllithium was reduced by reaction with DPE (in order to avoid reactions with the carbonyl groups), followed by the polymerization of the methacrylate monomer. Stearyl methacrylate, SMA is associated with

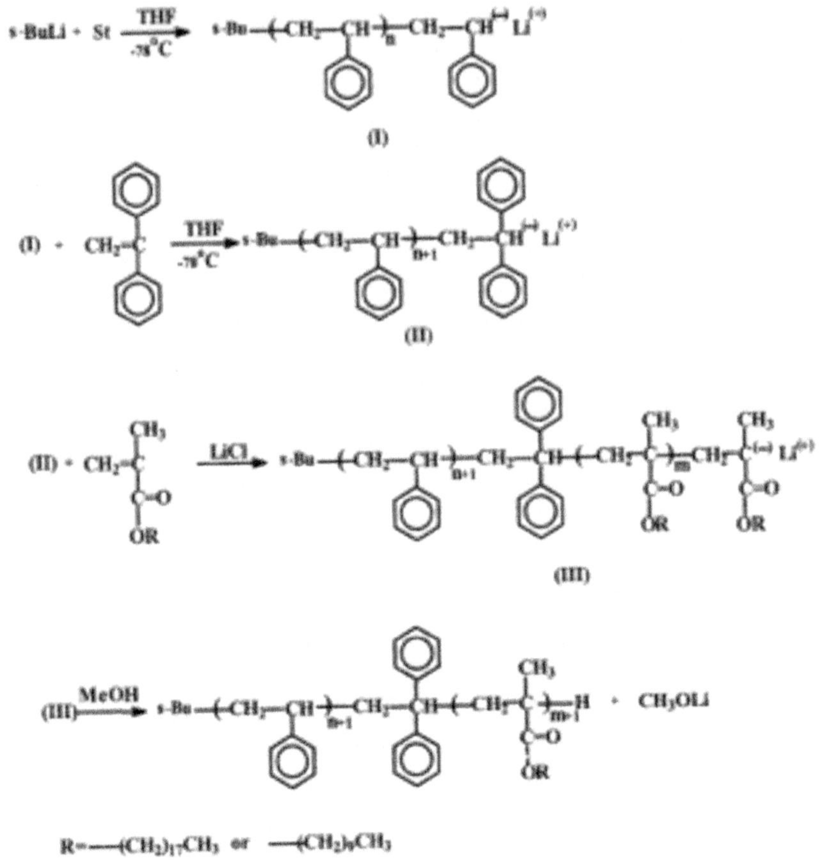

Scheme 1

two major limitations, the insolubility of the monomer in THF at – 78 °C and its very high boiling point, which prevents purification using standard methods developed for methacrylates [13]. To overcome these drawbacks the monomer was purified by recrystallization in n-hexane at – 30 °C, and the polymerization was conducted at – 10 °C, where SMA is soluble. Under these conditions living polymerization was promoted and led to well-defined block copolymers. This was confirmed by the efficient synthesis of a high molecular weight diblock copolymer and a triblock copolymer PS-*b*-PSMA-*b*-PS by coupling the living PS-*b*-PSMA diblocks with 1,4-dibromomethylbenzene as the linking agent. A combination of several characterization techniques revealed that model copolymers with high molecular and compositional homogeneity were prepared. The micellization properties of these samples were studied in solvents selective either for the PS or the polymethacrylate blocks.

Narrow molecular weight distribution PMMA-*b*-poly(2-perfluorooctylethyl methacrylate) block copolymers (Scheme 2) were synthesized in THF at

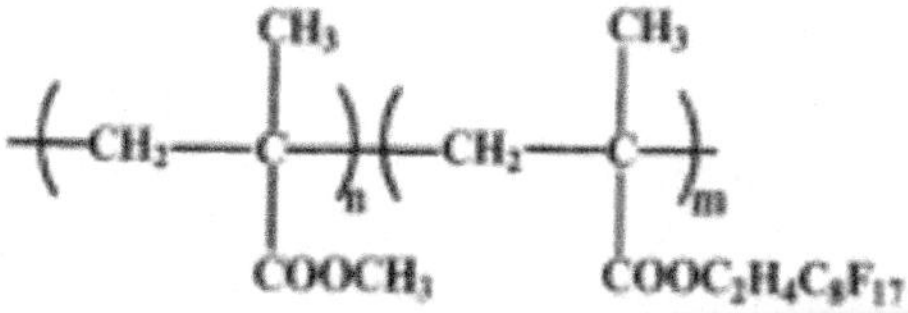

Scheme 2

– 78 °C in the presence of LiCl [14]. Their self-assembly behavior was studied in selective solvents. Diblock copolymers of *t*-butyl methacrylate, *t*BuMA, and 2-(*N*-methylperfluorobutanesulfonamido)ethyl methacrylate, FMA, were prepared by sequential addition of the monomers starting from *t*BuMA using 1,1-diphenyl-3-methylpentyllithium as the initiator [15]. Symmetric triblock copolymers PFMA-*b*-*t*BuMA-*b*-PFMA were synthesized using potassium naphthalene as a difunctional initiator to polymerize *t*BuMA, followed by the addition of FMA. The polymers were characterized by SEC and NMR spectroscopy, and their microphase separation behavior was studied by atomic force microscopy (AFM) and small angle X-ray scattering (SAXS). Thermal treatment of the copolymers at 200 °C yielded inter- and intramolecular anhydrides due to the splitting of the *t*-butyl ester groups associated with the formation of isobutene. Reflux in 1 N NaOH resulted in the formation of the sodium salt of the methacrylic acid, and consequently in the synthesis of amphiphilic copolymers (Scheme 3). Under these conditions the FMA units were not cleaved.

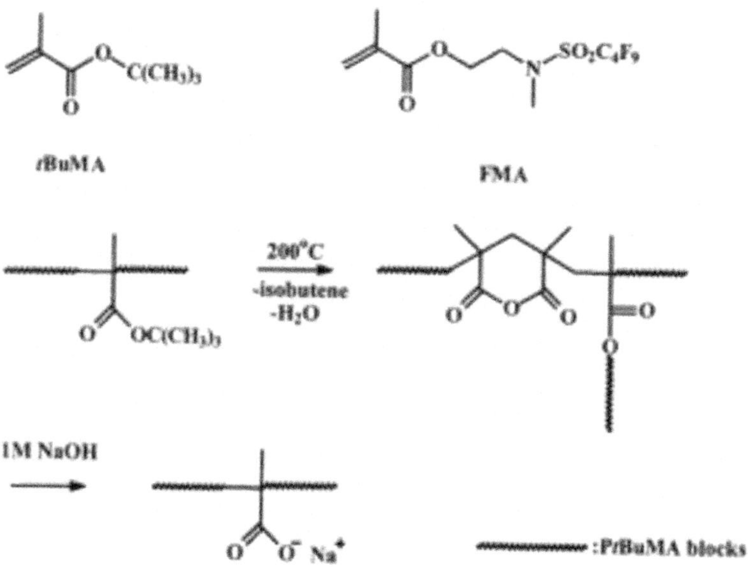

Scheme 3

Monofunctional or bifunctional low molecular weight poly(dimethylsiloxane), PDMS carrying one or two end-hydroxyl groups were used as macroinitiators for the synthesis of diblock and triblock copolymers, respectively, with 2-(dimethylamino)ethyl methacrylate, DMAEMA, PDMS-*b*-PDMAEMA, and PDMAEMA-*b*-PDMS-*b*-PDMAEMA [16]. The potassium salt of dimethyl sulfoxide, $DMSO^-K^+$ was used to convert the terminal hydroxyl groups of the PDMS chains to potassium alcoholates. Subsequent addition of DMAEMA led to the formation of the desired diblock or triblock copolymers. Extreme care should be given in the activation of the macroinitiator, especially regarding control of the stoichiometry of the reaction. Excess $DMSO^-K^+$ will act as an initiator to lead to the formation of PDMAEMA homopolymer along with the diblock or triblock copolymers. Smaller quantities of $DMSO^-K^+$ will lead to residual unactivated PDMS chains. Particularly, in the case of the bifunctional macroinitiator, deficient quantities of $DMSO^-K^+$ will provide a mixture of diblock, triblock, and PDMS homopolymer. Moderately broad molecular weight distributions ($M_w/M_n \sim 1.40$) were obtained. The micellar properties of these products were studied in selective solvents.

3-Methacryloyloxy-1,1′-biadamantane, MBA, (Scheme 4) was efficiently polymerized anionically using [1,1-bis(4′-trimethylsilylphenyl)-3-methylpentyl]lithium as the initiator, prepared *in situ* by the reaction of *s*-BuLi with 1,1-bis(4-trimethylsilylphenyl)ethylene [17]. The polymerization took place at − 50 °C in order to avoid the solubility problems of the monomer, observed at − 78 °C. Narrow molecular weight distribution block copolymers of rather low molecular weights of PMBA with *t*BuMA and (2,2-dimethyl-1,3-dioxolan-4-yl)methyl methacrylate were prepared. Using the difunctional potassium/naphthalene initiator triblock copolymers with polyisoprene middle blocks, PMBA-*b*-PI-*b*-PMBA were also synthesized.

Block copolymers with PS and a polymethacrylate block carrying a liquid crystalline group, PS-b-poly{6-[4-(cyanophenylazo)phenoxy]hexyl methacrylate}, were successfully prepared in quantitative yields and with relatively narrow molecular weight distributions (Scheme 5) [18]. The thermotropic liquid crystalline behavior of the copolymers was studied by differential scanning calorimetry.

A bifunctional methacrylate monomer, *trans*, *trans*-1-methacryloyloxy-2,4-hexadiene, MAHE, was efficiently polymerized anionically [19]. It was

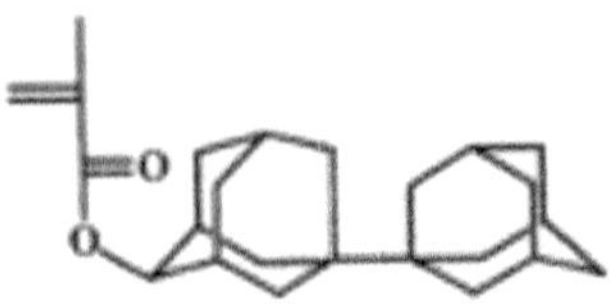

MBA

Scheme 4

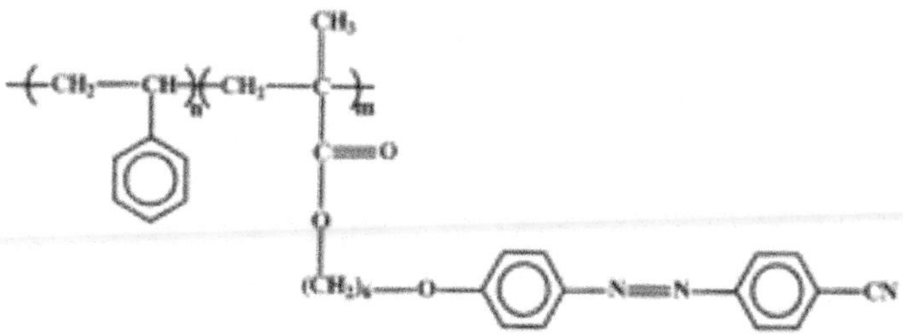

Scheme 5

found experimentally that the 2,4-hexadienyl side group was not affected during the polymerization. Block copolymers of MAHE with MMA were prepared by starting the polymerization with either monomer. Styrene was polymerized first to result in PS-*b*-PMAHE block copolymers. In all cases well-defined products were obtained. The pendant dienyl groups were further reacted with bromine or osmium tetroxide to generate amphiphilic functional block copolymers with bromide or hydroxyl side groups (Scheme 6).

Trialkylsilyl-protected oligo(ethylene glycol)methacrylates, 2-{2-[(*tert*-butyldimethylsilyl)oxy]ethoxy}ethyl methacrylate (1), and 2-{2-[2-[(*tert*-butyldimethylsilyl)oxy]ethoxy]ethoxy}ethyl methacrylate (2) (Scheme 7) were used for the synthesis of amphiphilic block copolymers by anionic poly-

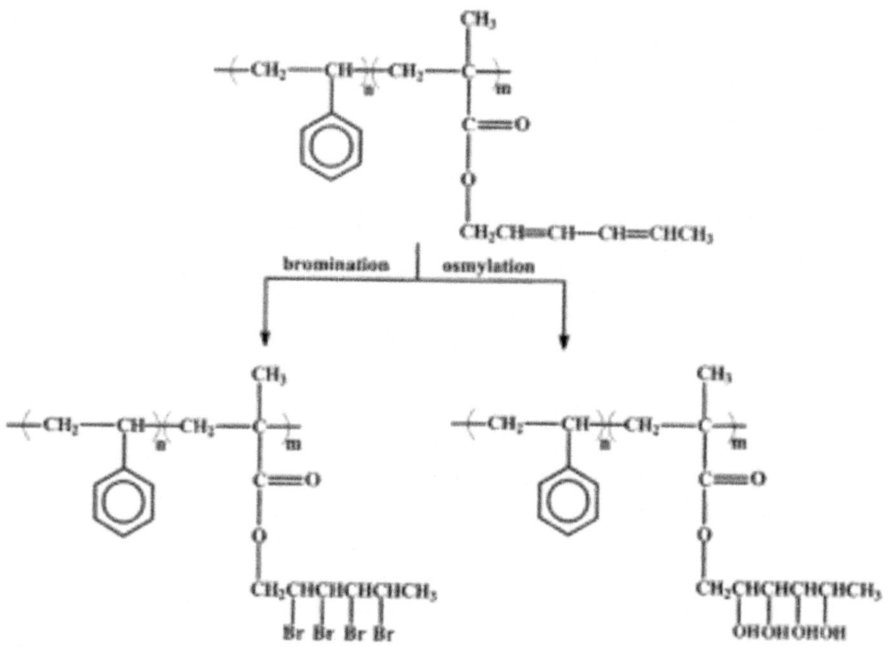

Scheme 6

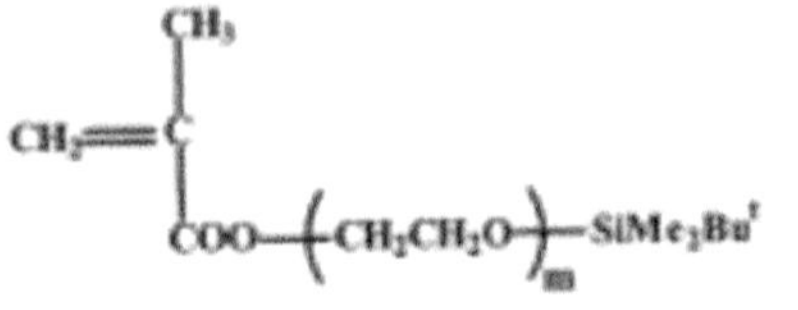

1: m=2
2: m=3

Scheme 7

merization [20]. The following structures were prepared: poly(1)-*b*-P*t*BuMA, poly(2)-*b*-P*t*BuMA, poly(1)-*b*-poly(2), PS-*b*-poly(1), and PS-*b*-poly(2). Deprotection of the trialkylsilyl groups was performed with 2N HCl in aqueous THF at 0 °C for 2 h yielding block copolymers containing the water-soluble poly[di(ethylene glycol) methacrylate] and poly[tri(ethylene glycol) methacrylate].

It is well known that the classic organolithium initiators fail to polymerize EO in weak or medium polarity solvents in the absence of suitable additives, due to the very strong lithium-oxygen bond produced during the initiation of EO, which is unable to promote the propagation reaction [21]. In contrast, benzyl potassium was successfully employed as an initiator for the synthesis of block copolymers containing ethylene oxide, EO. PS-*b*-PEO block copolymer, PI-*b*-P2VP-*b*-PEO triblock terpolymer, PS-*b*-PI-*b*-P2VP-*b*-PEO tetrablock quarterpolymer, and PS-*b*-PI-*b*-P2VP-*b*-P*t*BuMA-*b*-PEO pentablock quintopolymer were prepared by sequential addition of monomers [22, 23]. The monomer sequence was based on the relative nucleophilicity of the active centers. Detailed characterization data revealed that structures with predictable molecular weights, narrow molecular weight distribution, and chemical and compositional homogeneity were obtained in all cases. This work was further expanded to the synthesis of the diblock copolymer P2VP-*b*-PEO and the symmetric triblock copolymer P2VP-*b*-PEO-*b*-P2VP, given in Scheme 8 [24]. The latter structure was prepared by the reaction of the living P2VP-*b*-PEO diblock with p-dibromoxylene, as a linking agent. The linking reaction was performed in the presence of a catalytic amount of CsI, which transforms the $-CH_2Br$ groups to the more reactive $-CH_2I$ groups. Under these conditions linking was completed within 3 h instead of the 3 days without CsI. A small amount (~ 10%) of high molecular weight by-product was observed in most cases. This by-product was attributed to the reaction of the living ends with the pyridine ring to form the high molecular weight graft copolymer.

Further work related to the synthesis of copolymers with either P2VP or P4VP blocks has been reported in the literature. Triblock terpolymers PS-*b*-P2VP-*b*-PEO were synthesized in THF at – 78 °C by sequential polymerization of styrene and 2VP, initiated by *s*-BuLi in the presence of LiCl [25]. The living polymer was terminated with EO. The end-hydroxyl group was

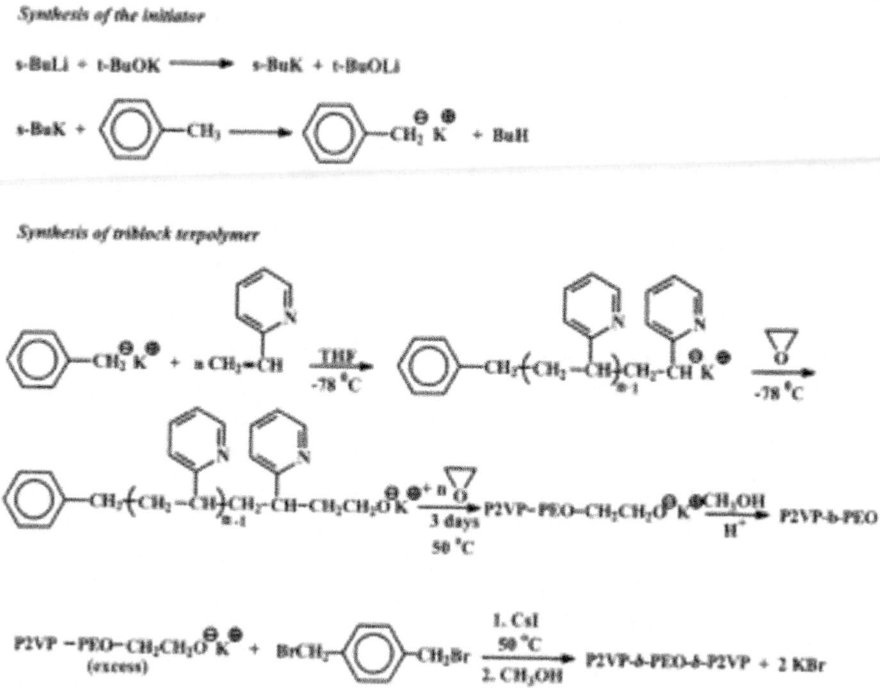

Scheme 8

treated with potassium naphthalene, and ethylene oxide was added and polymerized at 0 °C leading to a narrow molecular weight distribution product. The micellar properties of the terpolymer were studied in water.

Triblock terpolymers PS-*b*-PBd-*b*-P2VP and PBd-*b*-PS-*b*-P2VP, where PBd is polybutadiene (mostly 1,2-PBd), were prepared in order to study the microphase separation by transmission electron microscopy, TEM and SAXS. In the first case the triblocks were synthesized by the sequential addition of monomers in THF using *s*-BuLi as the initiator [26]. For the second type of copolymers, living PBd-*b*-PS diblocks were prepared in benzene at 40 °C in the presence of a small quantity of THF in order to obtain the desired 1,2-content and to accelerate the crossover reaction as well. DPE was then added to decrease the nucleophilicity of the active centers in order to avoid side reactions with the THF, which in combination with benzene was the solvent of the final step.

Symmetric triblock copolymers P4VP-*b*-PBd-*b*-P4VP were prepared using a difunctional initiator derived from the reaction of *m*-diisopropenylbenzene with *t*-butyllithium at – 20 °C (Scheme 9) [27]. The synthesis was conducted in a mixture of toluene and THF at temperatures higher than room temperature for the polymerization of Bd, followed by a lowering of the temperature at – 78 °C and finally addition of an extra quantity of THF and 4VP. The 4VP content was kept lower than 30% to avoid problems arising from the poor sol-

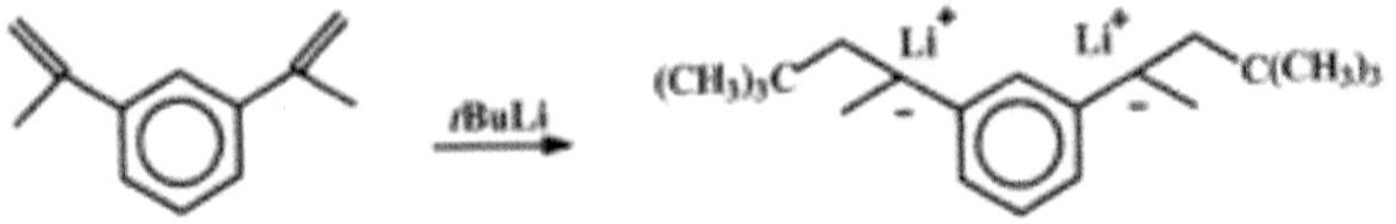

Scheme 9

ubility of the P4VP blocks in THF. Under these conditions chain branching side reactions were avoided.

A series of diblock AB and triblock ABA copolymers, where A is either P2VP or P4VP and B is poly(dimethylsiloxane), PDMS was prepared by anionic polymerization [28] (Scheme 10). Three different approaches were employed for this purpose. According to the first strategy (route A) an acetal-functionalized alkyllithium initiator was employed for the polymerization of 2VP or 4VP in THF at – 78 °C. Acid hydrolysis of the acetal group and titration of the hydroxyl groups with triphenyllithium gave a lithium alkoxide end group. These P2VP- and P4VP-lithium alkoxides can effectively initiate the polymerization of hexamethylcyclotrisiloxane, D_3 to produce the desired diblock copolymers. For extended polymerization times bimodal distributions were obtained, attributed to lithium silanolate aggregation. However, when high monomer concentrations (between 1.0 and 2.0 M) and relatively low monomer conversions (< 40%) were used, narrow molecular weight distribution products were obtained. Coupling of the lithium silanolate copolymers with dimethyldichlorosilane was effective in producing the ABA copolymers. According to the alternative route B, instead of using a functional initiator, a functional electrophilic termination reagent was employed. 2VP was polymerized with *t*BuLi in THF at – 78 °C. The living P2VP chains were terminated by the reaction with the suitable bromoacetal. The corresponding termination reaction of P4VP with the same bromoacetal was not effective, probably due to the lower reactivity of the P4VP anion that may be further complicated by the alkylation of the pyridine nitrogen with the bromoacetal. Following procedures similar to those in the previous method, the lithium alkoxide was prepared to give the diblock copolymer P2VP-*b*-PDMS or the triblock copolymer P2VP-*b*-PDMS-*b*-P2VP after coupling of the living diblock with dimethyldichlorosilane. In this case the conversion of the D_3 monomer was also kept low in order to avoid side reactions. The third method (route C) involves the end-capping of living P2VPLi with 2-isopropenylpyridine, IPP, followed by the addition of 1 equivalent of ethylene oxide. The lithium alkoxide, thus produced, efficiently promoted the polymerization of D_3 leading to the formation of the diblock and triblock copolymers. The effort to form the lithium alkoxide by the direct capping of P2VPLi with ethylene oxide finally afforded bimodal distributions. It was proposed that a picolyl carbanion [Scheme 10 (2)] is formed from the lithium alcoholate [Scheme 10 (1)]. This anionic site is not able to ini-

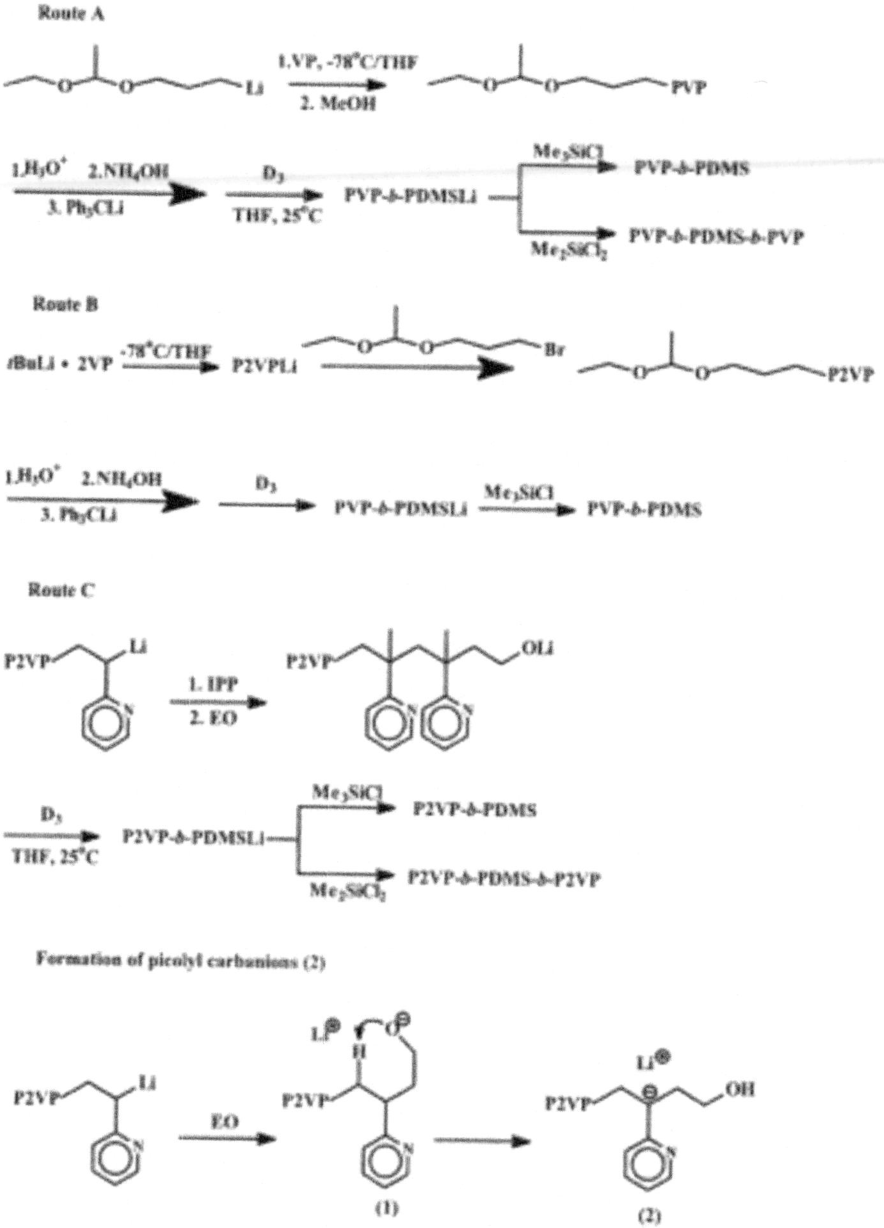

Scheme 10

tiate the polymerization of D_3, thus leading to bimodal distributions. The intramolecular rearrangement reaction was avoided by end-capping the living chains with 2-isopropenylpyridine lacking an α-hydrogen, responsible for the side reactions. Under these conditions well-defined copolymers were prepared. However, this end-capping reaction was not successful for P4VPLi chains.

The problems encountered for the polymerization of D_3 were successfully resolved by performing the polymerization in two steps [29]. In the first step the polymerization took place in benzene at room temperature with *n*-BuLi as the initiator. In a second step the temperature was lowered to − 20 °C, and the polymerization was allowed to proceed until completion after several days. With this methodology the PDMS was obtained in quantitative conversions, controlled molecular weights and narrow molecular weight distributions. Taking advantage of the living character of the polymerization, model triblock copolymers PS-*b*-PDMS-*b*-PS, were prepared by linking the living PS-*b*-PDMSLi with bis(dimethylchlorosilyl)ethane (Scheme 11). This specific linking agent was chosen to accelerate the coupling procedure and to avoid the side reactions at room temperature, where the linking reaction takes place.

1,3-Cyclohexadiene, CHD, is another monomer which presents a challenge for anionic polymerization due to the problems encountered for its controlled polymerization and its interesting properties both in solution and in bulk. Classic alkyllithium initiation leads to chain transfer reactions, as well as to the lithiated monomer, which is able to reinitiate polymerization. As a result, livingness of PCHD is difficult to achieve. It has been reported that living anionic polymerization of CHD can be achieved using *n*-BuLi and *N*,*N*,*N*′,*N*′-tetramethylethylenediamine, TMEDA, as the initiation system [30–32]. PCHD was found to have high thermal stability, low specific gravity, and good mechanical properties. Furthermore, the polymer can be transformed to polyphenylene (conductive polymer) by dehydrogenation and to polycyclohexylene (high Tg polymer) by hydrogenation.

Triblock copolymers PCHD-*b*-PBd-*b*-PCHD were also prepared by polymerizing Bd with a difunctional initiator, prepared by the reaction of 1,3-

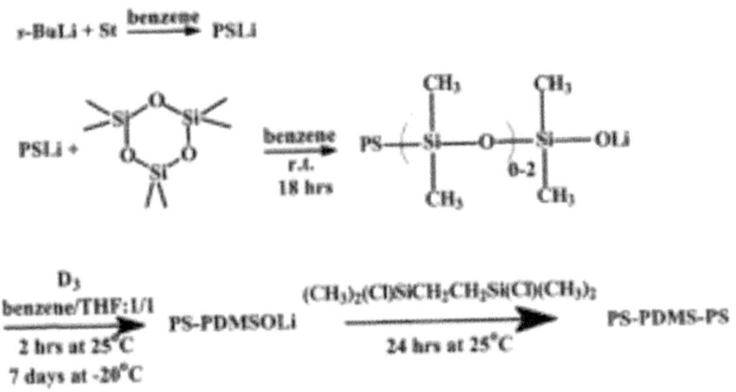

Scheme 11

di(1-propene-2-yl)benzene with *s*-BuLi in the presence of TMEDA, followed by the addition of CHD [33]. The reaction took place in cyclohexane at 40 °C, leading to well-defined products. The PBd block was subsequently hydrogenized selectively using the titanocene complex Cp_2TiCl_2 and diisobutylaluminum hydride in a molar ratio 1/6. Quantitative hydrogenation was conducted by heterogeneous catalysis using palladium supported on aluminum oxide.

It was found that PS-*b*-PCHD block copolymers can be prepared in hydrocarbon solvents using *s*-BuLi as the initiator without the presence of any additive [34]. Efficient crossover reactions were obtained from either PSLi (Scheme 12 route A) or PCHDLi (Scheme 12 route B). Using potassium/naphthalene as a difunctional initiator PS-*b*-PCHD-*b*-PS and PCHD-*b*-PS-*b*-PCHD were prepared. However, the molecular weight distributions were rather broad, and side reactions were observed when copolymers with high CHD contents were required. To avoid this problem several additives were tested in order to improve the copolymerization characteristics. The best results were obtained with dimethoxyethane, DME, or 1,4-diazabicyclo [4, 5] octane, DABCO, leading to narrow molecular weight distribution products. However, tailing effects or shoulders were observed in SEC chromatograms when the copolymers had CHD contents higher than 30%, meaning that the copolymer had to be purified by solvent/non-solvent fractionation.

Multiblock copolymeric structures containing PCHD blocks were also synthesized using *s*-BuLi as the initiator and either TMEDA or DABCO as the additive. Sequential monomer addition was performed with CHD being the last monomer added in all cases [35]. The structures prepared are: PS-*b*-PCHD, PI-*b*-PCHD and PBd-*b*-PCHD block copolymers, PS-*b*-PBd-*b*-PCHD, PBd-*b*-PS-*b*-PCHD and PBd-*b*-PI-*b*-PCHD triblock terpolymers, and PS-*b*-

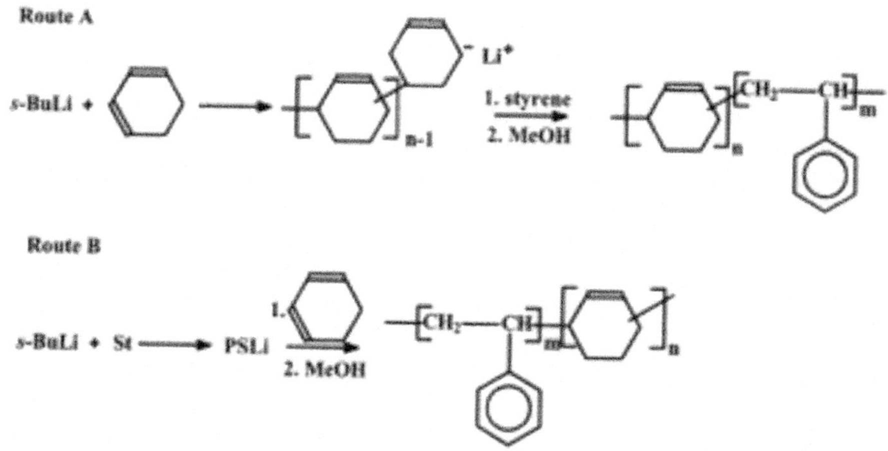

Scheme 12

PBd-*b*-PI-*b*-PCHD tetrablock quarterpolymers. In a few cases chain transfer or termination reactions led to the presence of a small amount of PCHD homopolymer. In general, detailed characterization revealed that narrow molecular weight distribution products with high chemical and compositional homogeneity were obtained.

(PS-*alt*-PCHD)-*b*-PS block copolymers were prepared in a one-pot procedure [36]. It was shown that the anionic statistical copolymerization of CHD and S in cyclohexane at 25 °C using *s*-BuLi as the initiator leads to an alternating copolymeric structure since the reactivity ratios for the two monomers were found to be 0.022 and 0.024 for CHD and S, respectively. When the molar ratio of S over CHD is greater than one, the desired (PS-*alt*-PCHD)-*b*-PS block copolymer is afforded. The molecular weights of the copolymers were rather low (around 10 000) and the molecular weight distributions narrow only when the S content was high. A gradual broadening of the distributions, by increasing the CHD content of the copolymers was observed. Subsequent selective hydrogenation of the CHD units using a nickel octoate catalyst and triethylaluminum cocatalyst yielded (PS-*alt*-polycyclohexane)-*b*-PS block copolymers in a quantitative manner. Aromatization of the CHD units using 2,3-dichloro-5,6-dicyano-1,4-benzoquinone under very mild conditions afforded (PS-*alt*-polyphenylene)-*b*-PS block copolymers.

Polyisocyanates represent a valuable class of polymeric materials since they adopt a rod-like helical conformation in solution and in bulk, and since they possess extraordinary liquid crystalline and optical properties [37–39]. Combination of these rigid chains with flexible blocks leads to novel materials characterized by the microphase separation between the different blocks along with the orientational ordering inside the rod-like phase. The synthesis of polyisocyanates by living anionic polymerization was hindered mainly by the existence of backbiting side reactions of the amidate anions leading to the formation of stable cyclic trimers [40]. The trimerization yield may reach 100% at temperatures higher than – 40 °C. Consequently, severe problems concerning the control over the molecular weight and the molecular weight distribution, the polymerization yield and the ability to prepare block copolymers or more complex architectures were exhibited. However, it was recently reported that the living anionic polymerization of isocyanates can be promoted when the polymerization is conducted at – 98 °C in THF with sodium naphthalene as the initiator in the presence of the crown ether 15-crown-5 or the salt sodium tetraphenylborate, $NaBPh_4$ [41, 42]. The crown ether entraps the sodium counter cations, thus depleting the anionic active center and greatly accelerating the propagation reaction. If the living polymer is terminated soon after the polymerization reaction is completed, the backbiting process is drastically reduced. On the other hand, $NaBPh_4$ stabilizes the amidate anion of the growing chain as a result of the common ion effect and the steric hindrance of the bulky tetraphenylborate group. $NaBPh_4$ was proven to be more efficient in promoting the living anionic polymerization of alkyl isocyanates. Using this methodology triblock copolymers composed of poly(hexyl isocyanate), PHIC, end blocks and either PS or PI middle blocks were prepared (simultaneously by the groups of Lee and Hadjichristidis [43–45]), using sodium/naphthalene as a difunctional initiator and $NaBPh_4$ as the additive. Furthermore, the Hadjichristidis' group reported the synthesis of pentablock terpolymers PHIC-*b*-PS-*b*-PI-*b*-PS-*b*-PHIC and PHIC-*b*-PI-*b*-PS-*b*-PI-*b*-PHIC (Scheme 13). Initial results concerning the microphase separation and the thermal properties of these materials were reported in these studies.

Hydrogenation reactions are the most common post-polymerization procedures, allowing for the synthesis of structures that would have never been prepared otherwise in such a controlled manner. Combinations of polydiene blocks and blocks carrying aromatic rings, as in PS, offer the possibility for selective hydrogenation of the polydienes, which are more susceptible to hydrogenation. This can be performed using homogeneous catalytic systems, such as the widely used Wilkinson catalyst. Usually heterogeneous catalysis with transition metals, either supported on inorganic surfaces or not, are very reactive but not selective. Examples have already been previously reported. In addition, the synthesis of polystyrene-*b*-poly(ethylene-*alt*-propylene)-*b*-polyethylene, PS-*b*-PEP-*b*-PE triblock terpolymers and PS-*b*-PEP-*b*-PS triblock copolymers by the homogeneous catalytic hydrogenation of the PS-*b*-PI-*b*-PBd and PS-*b*-PI-*b*-PS precursors, respectively, has been reported [46]. The reaction was conducted in toluene at 100 °C and 90 bar H_2 pressure with the Wilkinson catalyst. At lower temperatures the hydrogenation of the dienes was not quantitative.

PBd-*b*-PI-*b*-PEO triblock terpolymers were prepared by sequential addition of monomers using *s*-BuLi as the initiator [47]. The strong phosphazene base $tBuP_4$ was employed to promote the polymerization of ethylene oxide in the presence of a lithium counterion. Subsequent hydrogenation in toluene with the Wilkinson catalyst afforded the PE-*b*-PEP-*b*-PEO triblock terpolymers. Using *p*-toluenesulfonyl hydrazide as an alternative hydrogenation means, it was found that the PBd block was quantitatively hydrogenized, whereas the degree of hydrogenation was only 70% for the PI block, due to the steric hindrance involved in the reaction.

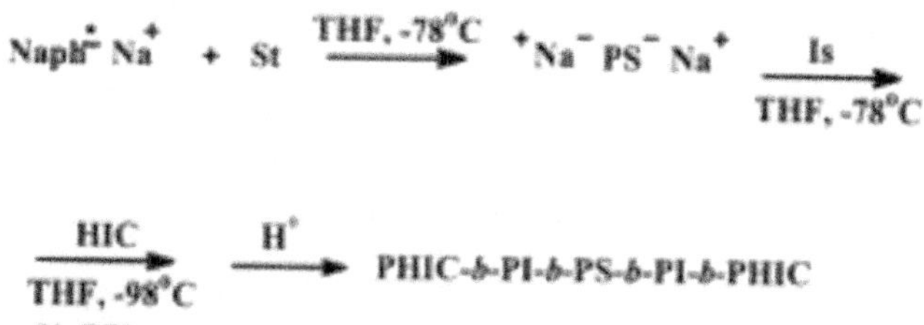

Scheme 13

2.3
Synthesis of Block Copolymers by Cationic Polymerization

Cationic polymerization was considered for many years to be the less appropriate polymerization method for the synthesis of polymers with controlled molecular weights and narrow molecular weight distributions. This behavior was attributed to the inherent instability of the carbocations, which are susceptible to chain transfer, isomerization, and termination reactions [48–52]. The most frequent procedure is the elimination of the cation's β-proton, which is acidic due to the vicinal positive charge. However, during the last twenty years novel initiation systems have been developed to promote the living cationic polymerization of a wide variety of monomers.

All the available new methods are aimed at stabilizing the carbocation by decreasing the positive charge of the growing cation, thereby reducing the acidity of the β-proton and eventually suppressing the chain transfer side reactions. Three methodologies (Scheme 14) were developed for this purpose: (1) *Protonic acid initiator and a mild Lewis acid*: The protonic acid initiator, e.g. HCl, forms an adduct with a dormant carbon-chlorine bond, which is electrophilically activated by the weak Lewis acid, such as metal halides, in order to initiate the living propagation. Living polymerization is attributed to the nucleophilic interaction of the carbocationic growing end with the binary counteranion. This method is usually applied successfully in non-polar solvents. (2) *Initiator, strong Lewis acid and Lewis base as the additive*: The use of a protonic initiator and a strong Lewis acid, e.g. $SnCl_4$, leads to poor control of the polymerization, due to the generation of binary counteranions, which are too weakly nucleophilic to efficiently stabilize the carbocations.

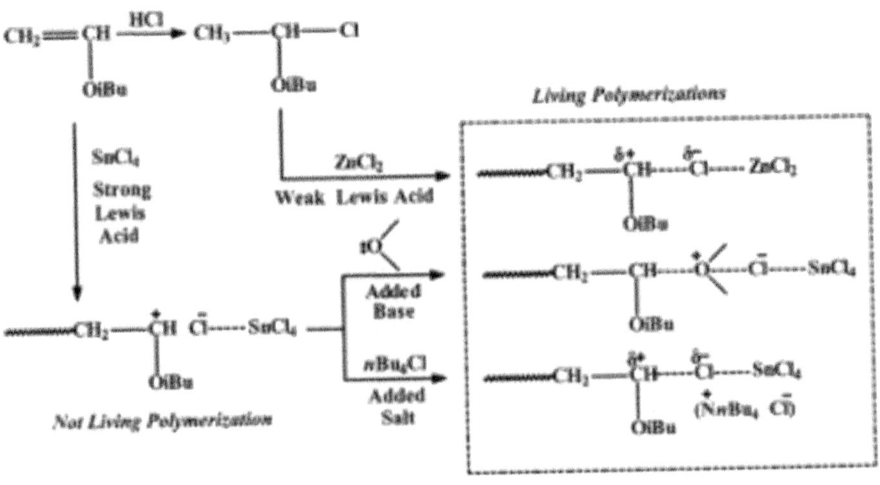

Scheme 14

However, in the presence of a suitable Lewis base the polymerization becomes living, due to the nucleophilic stabilization of the growing cation generated by the added base. (3) *Initiator, strong Lewis acid and onium salt as additive*: The previous method cannot be easily applied in polar media. In this case the living cationic polymerization is promoted by the addition of salts with nucleophilic anions, such as ammonium and phosphonium derivatives.

Applying these methodologies monomers such as isobutylene, vinyl ethers, styrene and styrenic derivatives, oxazolines, *N*-vinyl carbazole, etc. can be efficiently polymerized leading to well-defined structures. Compared to anionic polymerization cationic polymerization requires less demanding experimental conditions and can be applied at room temperature or higher in many cases, and a wide variety of monomers with pendant functional groups can be used. Despite the recent developments in cationic polymerization the method cannot be used with the same success for the synthesis of well-defined complex copolymeric architectures.

Block copolymers of methyl vinyl ether, MVE, and isobutyl vinyl ether, IBVE, of the type PMVE-*b*-PIBVE, PMVE-*b*-PIBVE-*b*-PMVE, and PIBVE-*b*-PMVE-*b*-PIBVE were prepared by living cationic polymerization and sequential monomer addition [53]. The acetal/trimethylsilyl iodide as the initiator and a ZnI_2 activator system was employed in all cases. 1,1-Diethoxyethane was the monofunctional initiator for the synthesis of the diblock copolymers, and 1,1,3,3-tetramethoxypropane was used as a difunctional initiator for the synthesis of the triblock copolymers (Scheme 15). Quantitative conversions were obtained for each monomer species. Rather narrow molecular weight distributions ($M_w/M_n < 1.2$) were obtained for the diblocks, whereas broader distributions ($1.25 < M_w/M_n < 1.30$) were observed for the triblocks, indicating that hydrogen impurities, due to the presence of water or hydrogen iodide, may exist in the polymerization system. These impurities may act as monofunctional initiators simultaneously with the difunctional initiator, thus leading to a mixture of triblock and diblock copolymers after the addition of the second monomer. Another explanation is that some of the active centers of the difunctional initiator may be deactivated by the presence of basic impurities, thus leading to a mixture of difunctional and monofunctional initiators. An additional possibility is associated with a common problem of the difunctional initiators, the different rate of initiation from the two active sites. For better control of the molecular weight distribution, the more reactive IBVE should be polymerized first.

Amphiphilic block copolymers consisting of poly(2-hydroxyethyl vinyl ether), POHVE, and poly(vinyl ethers) carrying fluorinated alkyl groups were prepared by sequential addition of monomers (Scheme 16) [54]. PHOVE was derived from the polymerization of acetoxy vinyl ether after hydrolysis of the acetoxy masking groups with sodium hydroxide in 1,4-dioxane. The electron withdrawing fluoroalkyl groups were introduced far away from the vinyl moieties, without affecting the reactivity of the active center during the poly-

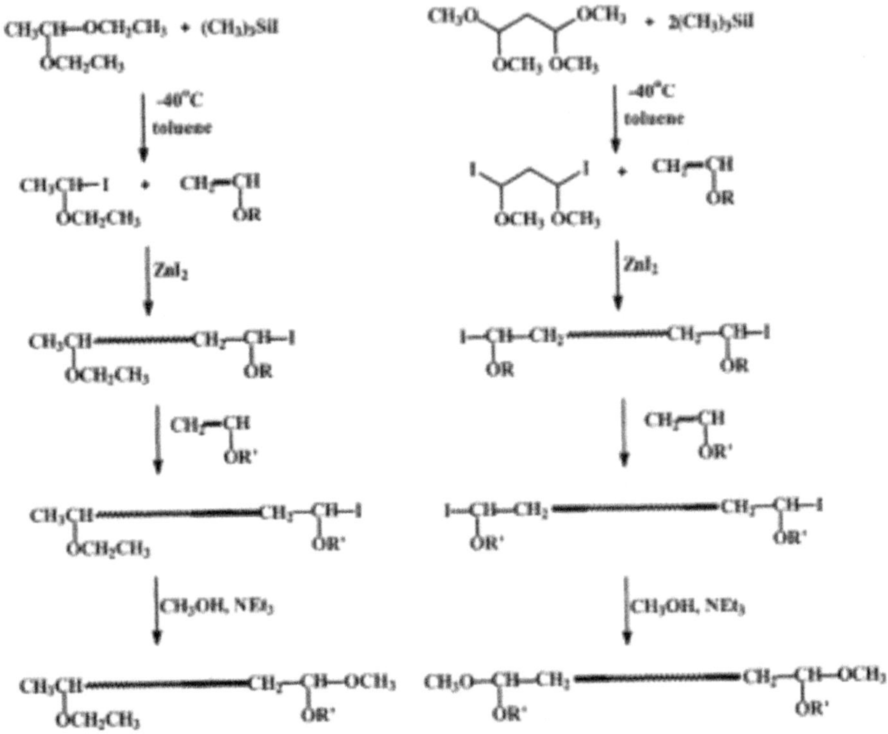

Scheme 15

merization. Narrow molecular weight distribution samples were obtained through this procedure. The micellar properties of these materials were investigated in aqueous solutions.

The 2-(*p*-methoxyphenyl)-ethanol-BF_3OEt_2, $CH_3CH(C_6H_4$-*p*-$OCH_3)OH$ – BF_3OEt_2 initiation system was successfully employed for the polymerization of *p*-alkoxystyrenes and *p*-hydroxystyrene, PHOS, without having to protect the hydroxyl groups to afford products with moderately broad molecular weight distributions ($M_w/M_n = 1.3$) [55]. The same initiation system was also employed for the synthesis of PHOS and poly(*p*-methoxystyrene), PMOS, starting from the polymerization of the former monomer in the presence of water in acetonitrile at 0 °C. After the consumption of the first monomer MOS and CH_2Cl_2 were added to afford the desired block copolymer. A rather broad molecular weight distribution ($M_w/M_n = 1.42$) and an asymmetric SEC chromatogram indicated that there was not absolute control during the copolymerization.

The cationic ring opening polymerization of ε-caprolactone, CL, and δ-valerolactone, VL, was investigated using *n*-BuOH/HCl·Et_2O as the initiation system [56]. It was observed that narrow molecular weight distribution samples were obtained. These results were combined with those previously

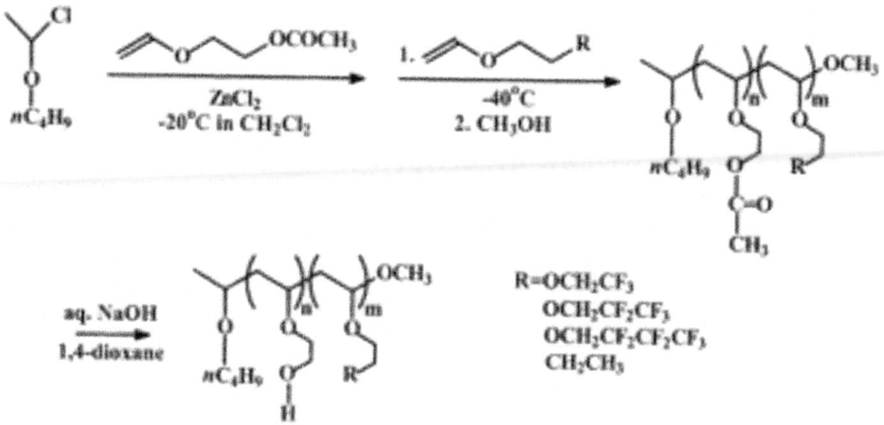

Scheme 16

reported for the polymerization of 1,3-dioxepan-2-one, 7CC, in order to prepare P(7CC)-*b*-PCL and P(7CC)-*b*-PVL (Scheme 17) diblock copolymers, starting from the polymerization of 7CC. Quantitative yields and products of narrow molecular weight distribution were obtained. Triblock terpolymers P(7CC)-*b*-PCL-*b*-PVL were also prepared using $H_2O/HCl{\cdot}Et_2O$ as the initiation system. It was found that the initiating carbonic acid polymer end of the P(7CC) smoothly changed into a hydroxyl group by rapid decarboxylation to form an α,ω-dihydroxyl polycarbonate. The procedure was proven to be efficient in producing narrow molecular weight distribution samples. However, only low molecular weight polymers were prepared.

Well-defined phosphazene block copolymers were prepared by the cationic polymerization of phosphoranimines [57]. Block copolymers of the type $[N=PCl_2]_n[N=PR(R')]_m$ were prepared using a wide variety of phos-

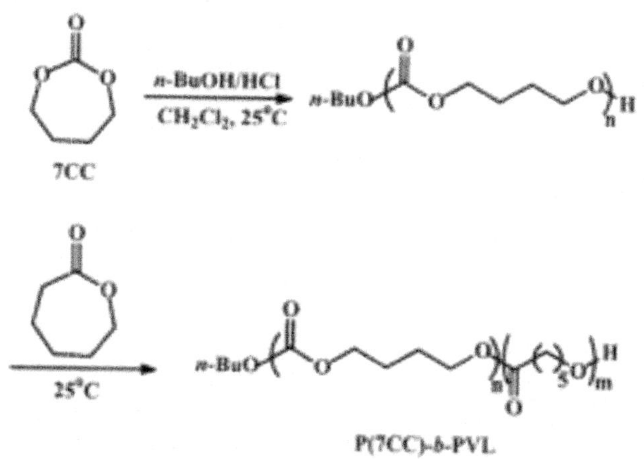

Scheme 17

phoronimines ($PhCl_2P=NSiMe_3$, $Me(Et)ClP=NSiMe_3$, $Me_2ClP=NSiMe_3$, $Ph_2ClP=NSiMe_3$, $PhF_2P=NSiMe_3$). PCl_5 was used as the initiator to polymerize $Cl_3P=NSiMe_3$ in CH_2Cl_2 at 35 °C as the first block, followed by the addition of the second monomer (Scheme 18). A living difunctional initiator was also used for the synthesis of ABA type triblock copolymers, as illustrated in Scheme 19.

Symmetric triblock copolymers of the ABA type, where B was PTHF and A poly(2-methyl-2-oxazoline), PMeOx, were prepared by cationic polymerization with trifluoromethanesulfonic anhydride as a difunctional initiator [58]. Subsequent hydrolysis of the PMeOx blocks with HCl in a methanol/water mixture resulted in the formation of the corresponding polyethylenimine blocks (Scheme 20). Samples with relatively low molecular weight distributions were obtained.

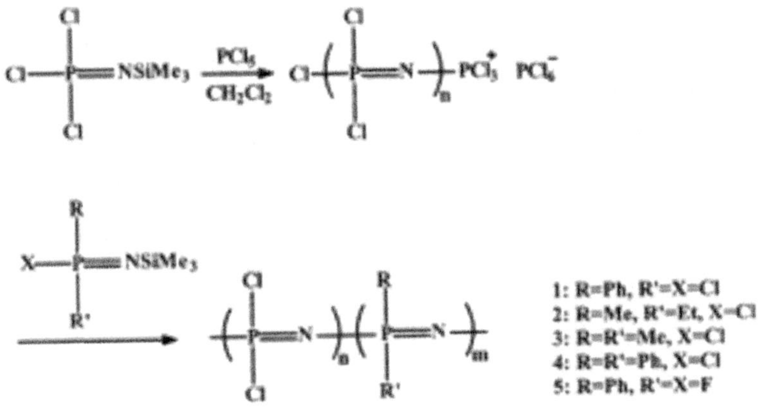

Scheme 18

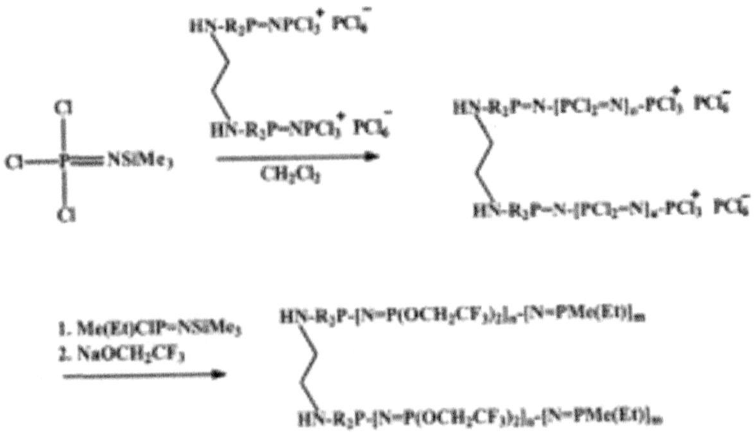

Scheme 19

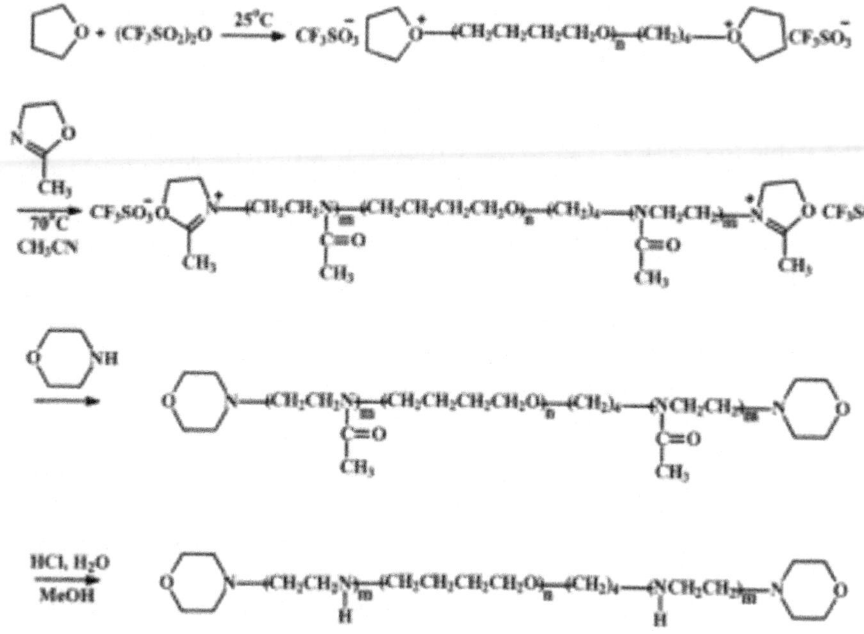

Scheme 20

2.4
Synthesis of Block Copolymers by Controlled Radical Polymerization

Free radical polymerization remains the most versatile method for polymer synthesis due to its compatibility with a wide range of monomers, including functional groups, its resistance to protic or aqueous media, allowing for the development of emulsion and suspension polymerization processes and to the experimentally less demanding conditions [59]. However, major limitations are associated with free radical polymerization including the lack of control of molecular weight, broad molecular weight distributions, and the inability to prepare complex macromolecular architectures. These drawbacks are due to the existence of several inherent termination reactions, such as disproportionation and chain transfer reactions.

Living free radical polymerization, combining the advantages of radical polymerization and those associated with the living polymerization technique, has long been recognized as the ideal situation in polymer chemistry. The common key to the control of radical polymerization lies in the reversible and rapid formation of dormant species. Under these equilibrium conditions, the instantaneous concentration of the active radical species is reduced, thus suppressing the termination reactions between the growing radicals. Recent advances in this area have greatly contributed to the realization of this concept. The first step in this direction involved the use of stable free radicals, such as nitroxides, as reversible termination agents to reduce termination reactions [60,61]. In the beginning, this nitroxide-mediated radical polymerization, NMP, was initiated by a bimolecular initiation system, consisting of a classical radical initiator, (e.g. benzoyl peroxide) and an alkoxyamine as the stable free radical [e.g. 2,2,6,6-tetramethyl-1-piperidinyloxy (TEMPO) radical]. By conducting the polymerization in bulk at elevated temperatures a benzyloxy radical is formed and subsequently undergoes reaction with monomer molecules to give a growing polymer chain. Reversible termination of this growing macromolecular chain with TEMPO leads to controlled growth and lower polydispersities than those obtained in free radical polymerization. However, this technique is not efficient during initiation and a variety of unwanted side reactions occur leading to poor control over the molecular characteristics. To overcome these complications unimolecular initiators were developed containing both the initiating radical and the alkoxyamine counter radical in the same molecule. The C – O bond is thermolytically unstable and decomposes on heating at elevated temperatures to give the initial radical and the mediating nitroxide radical. Recent advances in the synthesis of novel nitroxides with higher equilibrium constants for the cleavage of the corresponding alkoxyamines allowed for the application of the technique to a wide variety of monomers, such as styrene, methacrylates, acrylates, acrylamides, dienes, and acrylonitrile. The versatile nature of these initiators can also be used to control the synthesis of block copolymers from a wide selection of monomers. This technique, however, has several limitations. The upper molecular weight limit, for which there is agreement with the stoichiometric values, is between 150 000 and 200 000. The molecular weight distributions are generally lower than 1.20 but become broader upon increasing molecular weight, indicating the presence of termination reactions.

Novel catalytic systems, initially used for atom transfer radical additions in organic chemistry, have been employed in polymer science and referred to as atom transfer radical polymerization, ATRP [62–65]. Among the different systems developed, two have been widely used. The first involves the use of ruthenium catalysts [e.g. $RuCl_2(PPh_3)_2$] in the presence of CCl_4 as the initiator and aluminum alkoxides as the activators. The second employs the catalytic system CuX/bpy (X = halogen) in the presence of alkyl halides as the initiators. Bpy is a 4,4′-dialkyl-substituted bipyridine, which acts as the catalyst's ligand.

The general mechanism is given in Scheme 21. Activation of the organic halide R-X occurs via an electron transfer reaction between the transition metal (Mt^n-Y/ligand) and the organic halide (rate constant k_{act}), resulting in the formation of a radical. In this complex the transition metal oxidation number increases by one and the halide is covalently bound to the metal (X-Mt^{n+1}-Y/ligand). The resulting radical, R· then initiates the polymerization of the monomer (propagation rate constant k_p). During the propaga-

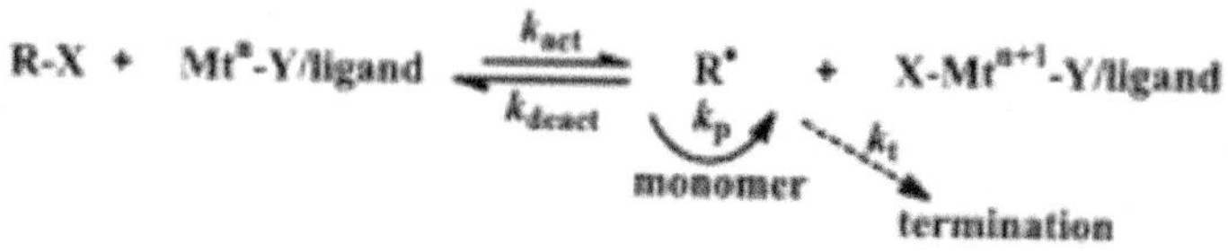

Scheme 21

tion the macromolecular chain reacts with the metal halide, which should be a deactivator of radical polymerizations (rate constant k_{deact}), to reform the lower oxidation state metal complex and a polymer chain with a halogen end group. The reaction repeats itself, using the polymer as the organic halide to reinitiate the polymerization. Termination reactions do occur in ATRP (rate constant k_t), mainly through radical coupling or disproportionation. There is no doubt that ATRP is among the most rapidly developing areas in polymer science, and makes possible the synthesis of well-defined macromolecules of complex architectures. As in the case of NMP, ATRP was initially applied for methacrylates and has a rather limited success with other monomers.

The reversible addition-fragmentation chain transfer (RAFT) method is another technique of controlled radical polymerization, based on the principle of degenerative chain transfer [66]. The process involves the conventional radical polymerization of a monomer in the presence of a chain transfer agent, CTA. The CTA usually contains a thiocarbonylthio group [S = C(– S – R)(– Z)] with proper substituents – *R* and – *Z* that influence the reaction kinetics and the macromolecular structural control. As shown in Scheme 22, a conventional radical initiator is used. However, in the presence of the CTA [Scheme 22, (1)] the polymerization does not proceed through the radicals formed by the initiator but from the initiating radicals $P_m^{\cdot}$. These radicals are produced from R· [Scheme 22, (4)], which is the fragmentation product of intermediate (2). This is achieved by utilizing a low concentration of initiator relative to CTA and a much higher reactivity of the CTA compared to the monomer. The equation ii (Scheme 22) including the consumption of CTA and reversible fragmentation of species (2) is usually referred to as pre-equilibrium, in order to differentiate from equation iv, which is the main equilibrium. The most important requirements for the production of polymers having controlled molecular weights and narrow molecular weight distributions are the following: (a) rapid establishment of the pre-equilibrium; (b) efficient re-initiation by the R· fragment; and (c) attainment of the main equilibrium in which the population of dormant chains and/or intermediate radicals [Scheme 22, (5)] (not reactive enough to add to monomers) is much higher than the total number of propagating chains $P_n^{\cdot}$ and $P_m^{\cdot}$. RAFT is an extremely versatile method regarding the monomer functionality and rigorous experimental techniques (vacuum line, inert atmosphere, use of extra pure reagents etc.) are not required.

Scheme 22

The controlled radical polymerization techniques opened up a new era in polymer synthesis, and further growth and developments are certain. However, the control of the molecular characteristics and the variety of macromolecular architectures reported by these methods cannot be compared with those obtained by other living polymerization techniques such as anionic polymerization.

2.4.1
Synthesis of Block Copolymers by Nitroxide-Mediated Radical Polymerization, NMP

Block copolymers consisting of a PS block and a poly(meth)acrylate or poly(vinyl acetate), PVAc, or poly(*N*,*N*, dimethylacrylamide), PDMA, block were prepared by NMP [67]. A variety of different side groups of the methacrylate were employed, such as methyl-, ethyl-, *n*-butyl-, *n*-octyl- and 2-(dimethylamino)ethyl acrylate. Styrene was polymerized first using the bimolecular initiator benzoyl peroxide and TEMPO. These PSs with TEMPO terminal groups served as macroinitiators for the polymerization of the second monomer. NMR and SEC analysis showed that the desired structures were prepared using this methodology. However, several drawbacks and limitations are associated with this technique. The molecular weight distributions were rather broad, leading to polydispersity indices up to 1.6 for both the PS homopolymers and the block copolymers. The conversions of the polymerizations although high were not quantitative and in some cases were rather low (lower than 30%). Furthermore, when the target molecular weight was higher than 10^5 there was always a deviation from the stoichiometric value, the experimental molecular weight being much higher.

The efficient polymerization of isoprene using 2,2,5-trimethyl-3-(1′-phenylethoxy)-4-phenyl-3-azahexane as a unimolecular initiator at 120 °C led to the synthesis of block copolymers carrying PI blocks [68]. P*t*BuA and PS macroinitiators were prepared with the previously mentioned initiator and used for the subsequent polymerization of isoprene, leading to the synthesis of the P*t*BuA-*b*-PI (Scheme 23) and PS-*b*-PI block copolymers. Rather narrow molecular weight distributions ($M_w/M_n < 1.2$) and high conversions (close to 80%) were observed. However, judging from the data presented for the homopolymerization of isoprene there is not very good agreement between the stoichiometric and the experimentally observed molecular weight. It was also shown that PI macroinitiators are not efficient for the polymerization of *t*BuA, but promotion of styrene polymerization is possible. Somewhat broader distributions were obtained for the PI-*b*-PS copolymers compared to the PS-*b*-PI samples.

The bulk polymerization of 2VP in the presence of TEMPO was successfully performed at 95 °C using acetic anhydride as an accelerator up to a conversion of 66% [69]. Side reactions were observed at higher conversions leading to broad molecular weight distributions. Taking advantage of these results block copolymers with styrene and *t*BuMA were prepared. Better control was observed when the PS content was higher than that of P2VP.

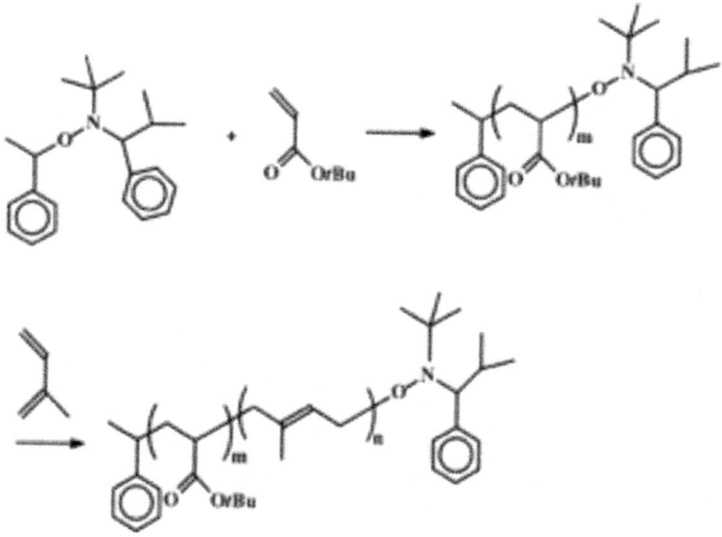

Scheme 23

Furthermore, the polydispersity increased by increasing the 2VP content. Similar results were obtained for the copolymers with *t*BuMA.

PS macroinitiators were prepared using the monofunctional and difunctional nitroxide initiators, shown in Scheme 24, leading to the synthesis of semitelechelic and telechelic chains [70]. These macroinitiators were used for the polymerization of 2,5-bis[(4-butylbenzoyl)oxy]styrene, BBOS, a styrene derivative with mesogenic side groups to produce PS-*b*-PBBOS block copolymers and PBBOS-*b*-PS-*b*-PBBOS triblock copolymers. The polymerization was conducted in *o*-dichlorobenzene using 50 wt % solutions, leading to polydispersities up to 1.3. It was also possible to directly dissolve the PS macroinitiators in a BBOS melt at 95 °C and perform the polymerization in bulk at 125 °C. The molecular weight distributions were broader in this case (up to

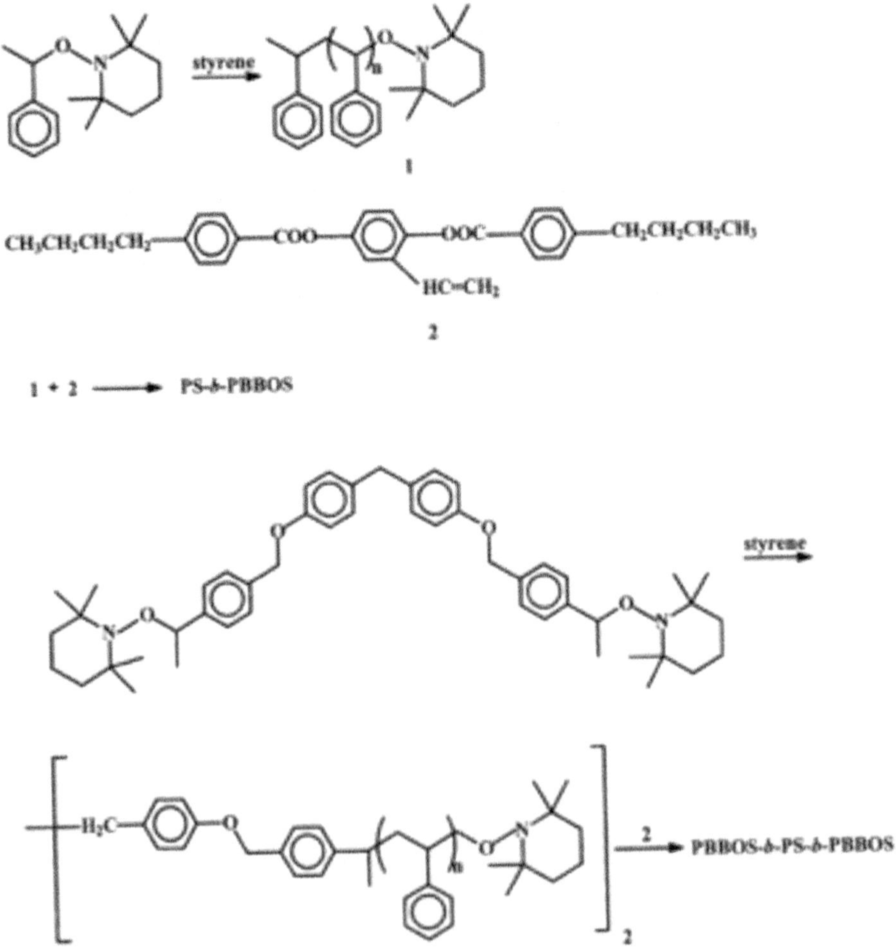

Scheme 24

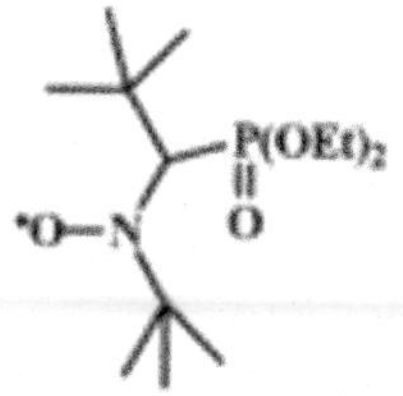

Scheme 25

1.45). The microphase separation and the rheological properties of these materials were subsequently studied.

A *β*-phosphonylated nitroxide, *N*-*t*-butyl-*N*-(1-diethylphosphono-2,2-dimethylpropyl)nitroxide, DEPN, (Scheme 25) in combination with AIBN was employed as a bimolecular initiator for the polymerization of 4VP and *N*,*N*-dimethylacrylamide, DMAA, in bulk at 110 °C [71]. The molecular weight distributions were rather narrow for P4VP (up to 1.3) but broader for PDMAA (higher than 1.3). The conversions were rather low, but in the case of the polymerization of DMAA were much higher than previous results reported using TEMPO rather than DEPN. Efforts to synthesize PDMAA-*b*-P4VP block copolymers failed to give copolymers with DMAA contents higher than 40%. When P4VP macroinitiators were employed for the polymerization of DMAA better results were obtained with polydispersities lower than 1.4 and higher DMAA contents. However, there was not good agreement between the stoichiometric and the experimentally observed molecular weights, especially at the higher molecular weights.

2.4.2
Synthesis of Block Copolymers by Atom Transfer Radical Polymerization, ATRP

The polymerization of 2-(diethylamino)ethyl methacrylate, DEAEMA, was studied under different conditions. It was shown that the best system providing narrow molecular weight distribution polymers involved the use of *p*-toluenesulfonyl chloride/CuCl/HMTETA as the initiator/catalyst/ligand at 60 °C in methanol [72]. Taking advantage of these results, well-defined PDEAEMA-*b*-P*t*BuMA block copolymers were obtained. The synthesis was successful when either *t*BuMA or DEAEMA was polymerized first. Poor results with bimodal distributions were obtained when CuBr was used as the catalyst. This behavior was attributed to the poor blocking efficiency of PDEAEMA-Br and the incomplete functionalization of the macroinitiator.

Monofunctional 2-bromo-2-methylpropionic acid 4-methoxyphenyl ester and difunctional 1,4-(2′-bromo-2′-methyl-propionate)benzene initiators, given in Scheme 26, were employed for the polymerization of *n*-BuMA followed by the addition of DMAEMA, thus leading to the formation of P*n*BuMA-*b*-PDMAEMA block and PDMAEMA-*b*-P*n*BuMA-*b*-PDMAEMA tri-

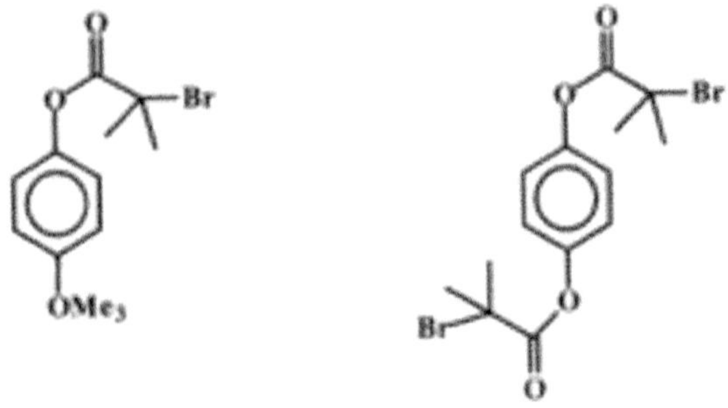

Scheme 26

block copolymers, respectively [73]. The synthesis took place in toluene using CuBr and N-(*n*-propyl)-2-pyridylmethanimine as the catalyst and the ligand, respectively. The polydispersity indices were in a range between 1.24 and 1.78. The polymers were quaternized with methyl iodide to render them more hydrophilic. The aggregation behavior of these materials was subsequently studied in aqueous solutions.

The ATRP of 4VP was efficiently performed using 1-phenylethyl chloride as the initiator and 5,5,7,12,12,14-hexamethyl-1,4,8,11-tetraazamacrocyclotetradecane as the ligand [74]. The polymerization was conducted in 2-propanol at 40 °C leading to almost quantitative yields and relatively narrow molecular weight distributions. The linear evolution of molecular weight with conversion, the constant concentration of the chain radicals during the polymerization and the controlled molecular characteristics strongly support the living character of the polymerization. With the P4VP-Cl macroinitiator the polymerization of styrene was conducted in DMF at 110 °C and the same catalyst/ligand system. Consequently, P4VP-*b*-PS block copolymers with low polydispersities were obtained.

ABC-type triblock terpolymers, where A was P*t*BuA, B was PS and C was poly(methyl acrylate), PMA, were synthesized by sequential addition of monomers using CuBr/PMDETA as the catalyst/ligand system and methyl-2-bromopropionate as the initiator [75]. SEC analysis showed that no polymer termination took place in the chain extension, resulting in narrow molecular weight distributions. More complex structures such as the pentablock terpolymers of the type ABCBA were also prepared using the difunctional initiator dimethyl-2,6-dibromoheptanedioate. The catalyst system CuBr/PMDETA was employed for the synthesis of PMA-*b*-P*t*BuA-*b*-PS-*b*-P*t*BuA-*b*-PMA terpolymers leading to well-defined structures. Bromo-terminated P*t*BuA-*b*-PS-*b*-P*t*BuA triblock were also used as macroinitiators for chain extension with MMA. To accelerate the crossover reaction CuCl was used to invoke the halogen exchange. Without this procedure the efficiency of the cross-propagation from the acrylate to the methacrylate chain end was poor. Furthermore, HMTETA was used as the complexing agent to avoid the heterogeneity associated with the corresponding PMDETA complex in MMA. Under these conditions well-defined PMMA-*b*-P*t*BuA-*b*-PS-*b*-P*t*BuA-*b*-PMMA pentablock terpolymers were prepared. The difunctional P*t*BuA macroinitiator was chain extended with MMA using CuCl/HMTETA, followed by further extension with 4VP using CuCl/tris-[(2-dimethylamino)ethyl]amine (Me_6TREN). A broader molecular weight distribution and an asymmetric SEC trace, revealing an overlap with the SEC trace of the triblock copolymer precursor indicated the occurrence of termination reactions during the addition of 4VP and/or problems concerning the cross-propagation reaction.

Several conditions were examined for the more efficient polymerization of MMA starting from the difunctional macroinitiator of P*n*BuA, prepared using the difunctional initiator diethyl *meso*-2,5-dibromoadipate and $NiBr_2(PPh_3)_2$ [76]. The best system involved the use of CuCl/dNbipy. As shown in the previous study the halogen exchange from bromine to chlorine was necessary to accelerate the crossover reaction from the acrylate to the methacrylate chain end. Addition of 10% $CuCl_2$ was efficient in reducing the polydispersity of the final products. The rheological properties of the triblock copolymers PMMA-*b*-P*n*BuA-*b*-PMMA were examined.

The direct synthesis of poly(3-sulfopropyl methacrylate)-*b*-PMMA, PSPMA-*b*-PMMA (Scheme 27) without the use of protecting chemistry, by sequential monomer addition and ATRP techniques was achieved [77]. A water/DMF 40/60 mixture was used to ensure the homogeneous polymerization of both monomers. CuCl/bipy was the catalytic system used, leading to quantitative conversion and narrow molecular weight distribution. In another approach the PSPMA macroinitiator was isolated by stopping the polymerization at a conversion of 83%. Then using a 40/60 water/DMF mixture MMA was polymerized to give the desired block copolymer. In this case no residual SPMA monomer was present before the polymerization of MMA. The micellar properties of these amphiphilic copolymers were examined.

A novel coil-rod-coil triblock copolymer, where the rod block was polyfluorene, PF, and the coil blocks poly(2-tetrahydropyranyl methacrylate),

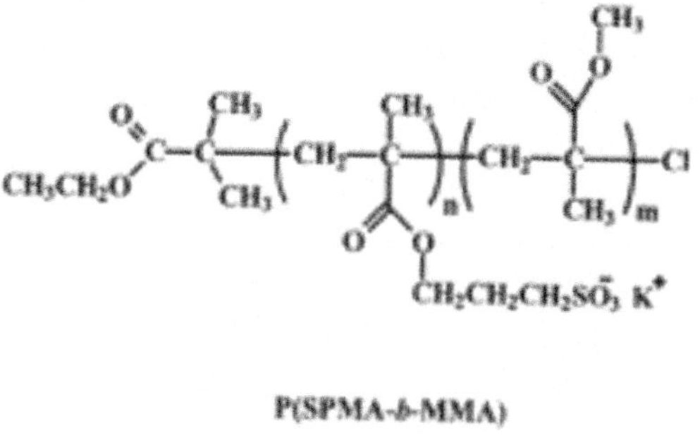

Scheme 27

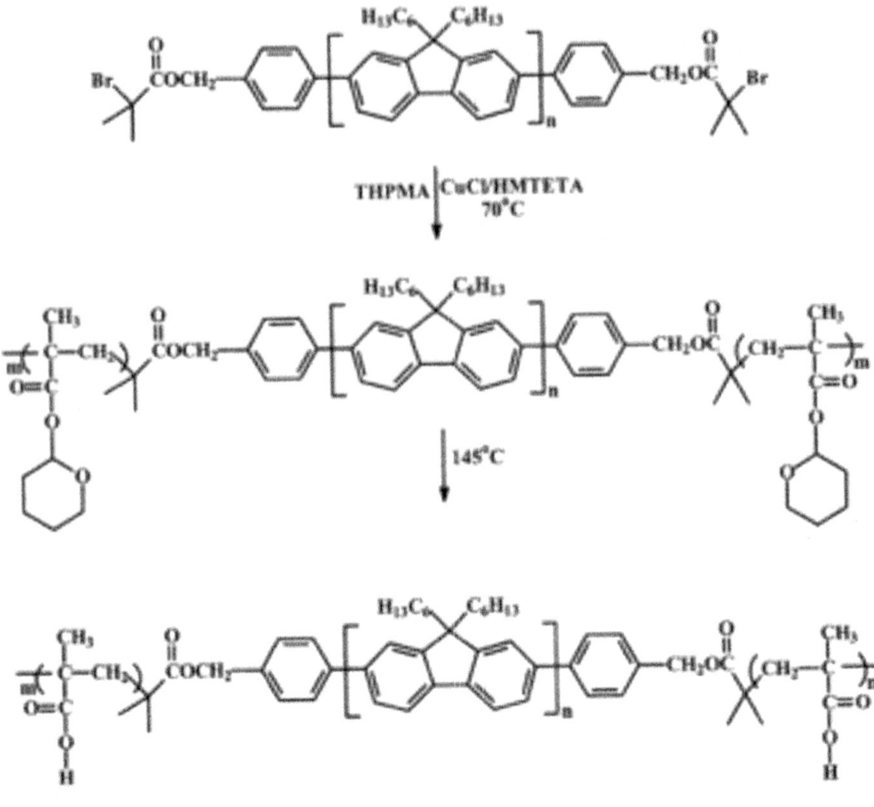

Scheme 28

PTHPMA, were prepared by ATRP techniques (Scheme 28) [78]. A well-defined PF difunctional initiator with 2-bromoisobutyrate end groups was employed for the polymerization of THPMA in *o*-dichlorobenzene at 60 °C or 70 °C using CuCl/HMTETA. As in previous studies, a halogen exchange was performed to accelerate the crossover reaction. Conversions higher than 80% and polydispersities lower than 1.3 were obtained. Thermolysis at 150 °C, in the presence of water, removed the tetrahydropyranyl protecting groups transforming the PTHPMA blocks to poly(methacrylic acid). Acid hydrolysis in the specific case could not be performed since chain scission or changes in the electronic structure of PF may occur.

A half-metallocene iron iodide carbonyl complex $Fe(Cp)I(CO)_2$ was found to induce the living radical polymerization of methyl acrylate and *t*-butyl acrylate with an iodide initiator $(CH_3)_2C(CO_2Et)I$ and $Al(Oi\text{–}Pr)_3$ to provide controlled molecular weights and rather low molecular weight distributions ($M_w/M_n < 1.2$) [79]. The living character of the polymerization was further tested with the synthesis of the PMA-*b*-PS and P*t*BuA-*b*-PS block copolymers. The procedure efficiently provided the desired block copolymers, albeit with low molecular weights.

2.4.3
Synthesis of Block Copolymers by Reversible Addition-Fragmentation Chain Transfer Radical Polymerization, RAFT

RAFT polymerization of two anionic acrylamido monomers: sodium 2-acrylamido-2-methylpropane-sulfonate, AMPS, and sodium 3-acrylamido-3-methyl-butanoate, AMBA, (Scheme 29) was conducted in water at 70 °C using 4,4′-azobis(4-cyanopentanoic acid) as the initiator and 4-cyanopentanoic acid dithiobenzoate as the RAFT chain transfer agent [80]. The synthesis was initiated either from one monomer or the other leading to narrow molecular weight distributions in both cases ($M_w/M_n < 1.2$).

Monofunctional and difunctional xanthates, shown in Scheme 30, were employed as chain transfer agents in the synthesis of block and triblock copolymers of acrylic acid, AA and acrylamide, AAm: PAA-*b*-PAAm, PAAm-*b*-PAA-*b*-PAAm and P(AA-*stat*-AAm)-*b*-PAAm [81]. The polymerizations were conducted in aqueous solutions at 70 °C with 4,4′-azobis(4-cyanopentanoic acid) as the initiator. The yields were almost quantitative,

AMPS AMBA

Scheme 29

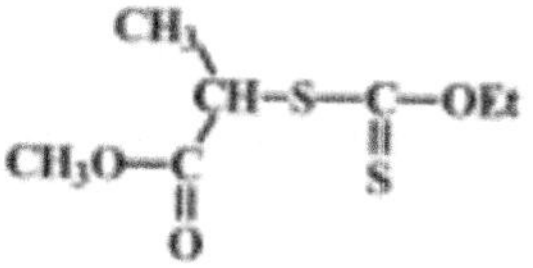

xanthate 1

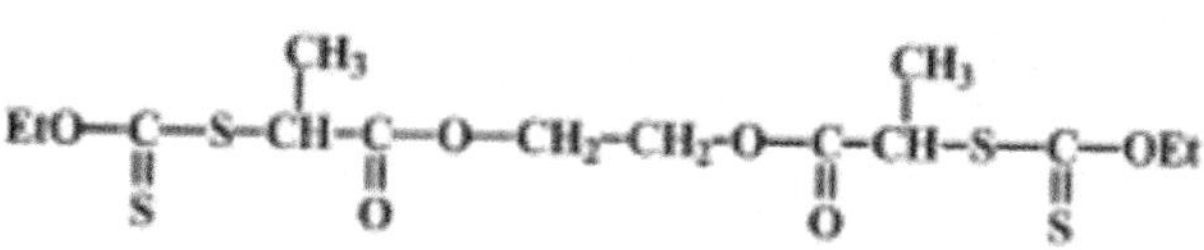

xanthate 2

Scheme 30

without very good control of the molecular weight in all cases, and with rather broad molecular weight distributions ($M_w/M_n > 1.3$).

Water soluble block copolymers consisting of *N*-isopropylacrylamide, NIPA, and the zwitterionic monomer 3-[*N*-(3-methacrylamidopropyl)-*N*,*N*-dimethyl]ammoniopropane sulfonate, SPP, were prepared via the RAFT process [82] (Scheme 31). NIPA was polymerized first using AIBN as the initiator and benzyl dithiobenzoate as the chain transfer agent. To avoid the problem of incomplete end group functionalization the polymerization yield was kept very low (less than 30%). The block copolymerization was then performed

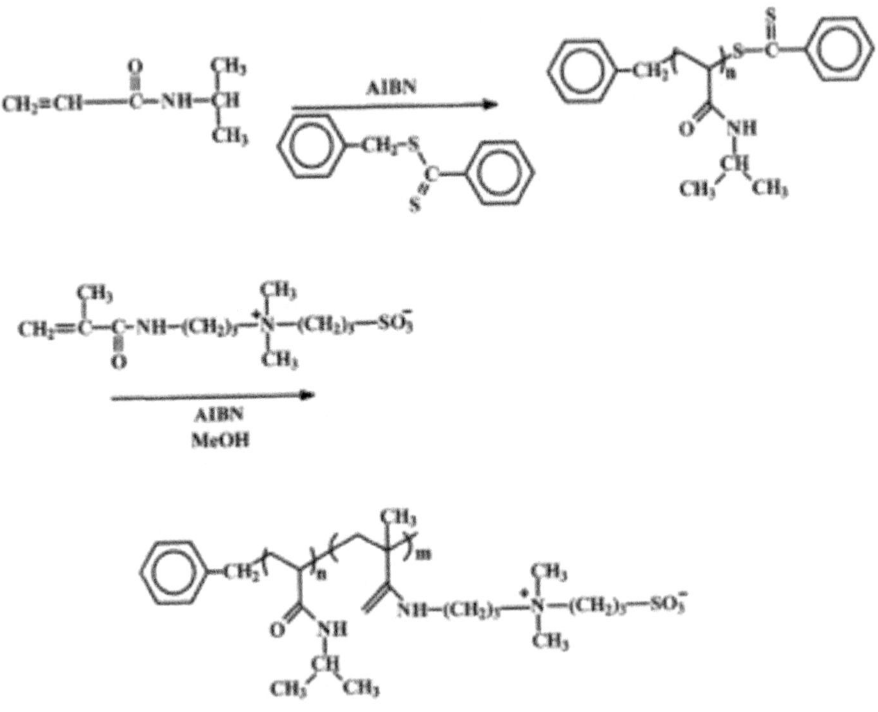

Scheme 31

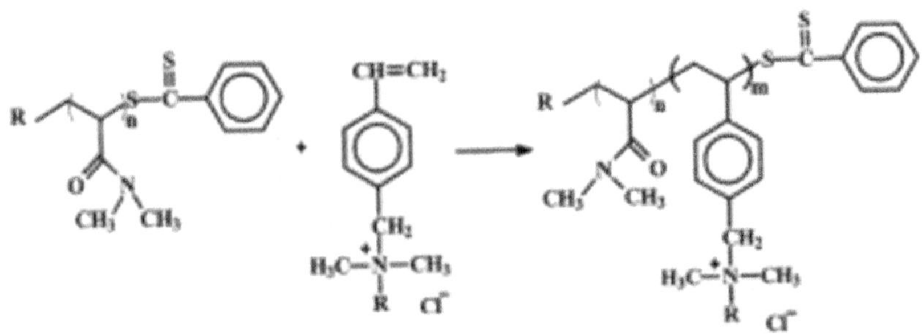

Scheme 32

in methanol. The conversion in this case as well was not quantitative, indicating that there is no control over the molecular weight. The molecular characterization of the samples was rather limited.

Block copolymers of *N,N*-dimethylacrylamide, DMAA, and *N,N*-dimethylvinylbenzylamine, DMVBAC, were prepared using 4,4′-azobis(4-cyanopentanoic acid) as the initiator and 4-cyanopentanoic acid dithiobenzoate as the chain transfer agent (Scheme 32) [83]. Starting the copolymerization procedure from DMVBAC a bimodal distribution was finally observed. Using the reverse order of addition, starting the copolymerization from DMAA followed by the addition of DMVBAC, well-defined structures were obtained. The low blocking efficiency of the PDMVBAC block may be attributed to preferential fragmentation of the intermediate radical formed during the preequilibrium of the copolymerization, which predominately yields the DMAA propagating chain.

2.5
Synthesis of Block Copolymers by Group Transfer Polymerization, GTP

The controlled polymerization of (meth)acrylates was achieved by anionic polymerization. However, special bulky initiators and very low temperatures (– 78 °C) must be employed in order to avoid side reactions. An alternative procedure for achieving the same results by conducting the polymerization at room temperature was proposed by Webster and Sogah [84]. The technique, called group transfer polymerization, involves a catalyzed silicon-mediated sequential Michael addition of α,β-unsaturated esters using silyl ketene acetals as initiators. Nucleophilic (anionic) or Lewis acid catalysts are necessary for the polymerization. Nucleophilic catalysts activate the initiator and are usually employed for the polymerization of methacrylates, whereas Lewis acids activate the monomer and are more suitable for the polymerization of acrylates [85, 86].

The method has been applied mainly for methacrylates and acrylates but other monomers, such as methacrylonitrile, acrylonitrile, dienoates etc., have been used as well. The polymerization is compatible with functional groups, i.e. dimethylamine-, glycidyl-, vinyl benzyl-, allyl- etc. However, groups bearing active hydrogens, such as hydroxyl, carboxylic acid, phenol, primary or secondary amines etc., interfere with the polymerization. A major limitation of the procedure is the molecular weight range for which the method is living. For molecular weights higher than 100 000 there is no agreement between the stoichiometric and the experimental value, the yields are not quantitative, and in addition broad molecular weight distributions are observed. Block copolymers can be easily prepared with this method, usually by sequential addition of monomers. In the case of copolymers consisting of a methacrylate and an acrylate block the polymerization of the methacrylate monomer is conducted first to give a well-defined product.

GTP was employed for the synthesis of block copolymers with the first block PDMAEMA and the second PDEAEMA, poly[2-(diisopropylamino)ethyl methacrylate], PDIPAEMA or poly[2-(*N*-morpholino)ethyl methacrylate], PMEMA (Scheme 33) [87]. The reactions took place under an inert atmosphere in THF at room temperature with 1-methoxy-1-trimethylsiloxy-2-methyl-1-propane, MTS, as the initiator and tetra-*n*-butyl ammonium bibenzoate, TBABB, as the catalyst. Little or no homopolymer contamination was evidenced by SEC analysis. Copolymers in high yields with controlled molecular weights and narrow molecular weight distributions were obtained in all cases. The micellar properties of these materials were studied in aqueous solutions.

Oligo(ethylene glycol) monomethyl ether monomethacrylate, OEGMA, was copolymerized with either benzyl methacrylate, BzMA or THPMA to afford the corresponding block copolymers via GTP [88]. MTS and TBABB were employed as initiator and catalyst, respectively. High yields and narrow molecular weight distributions were obtained in all cases. BzMA and THPMA were considered as precursors to obtain the MA residues. In the case of BzMA deprotection was attempted by hydrogenolysis. However, incomplete debenzylation of BzMA-rich copolymers and contamination of the final products with catalyst residues limited the utility of the procedure. In contrast THPMA-based copolymers could be deprotected efficiently by acid hydrolysis under mild conditions regardless of the block composition. The post-polymerization reactions are reported in Scheme 34. The aqueous solution properties of these samples were investigated.

ABC, ACB, and BAC triblock terpolymers, where A is PMMA, B is PDMAEMA, and C is poly[hexa(ethylene glycol)methacrylate], PHEGMA, were synthesized via GTP and sequential monomer addition [89]. The polymerizations were conducted in THF using MTS and TBABB as the initiator

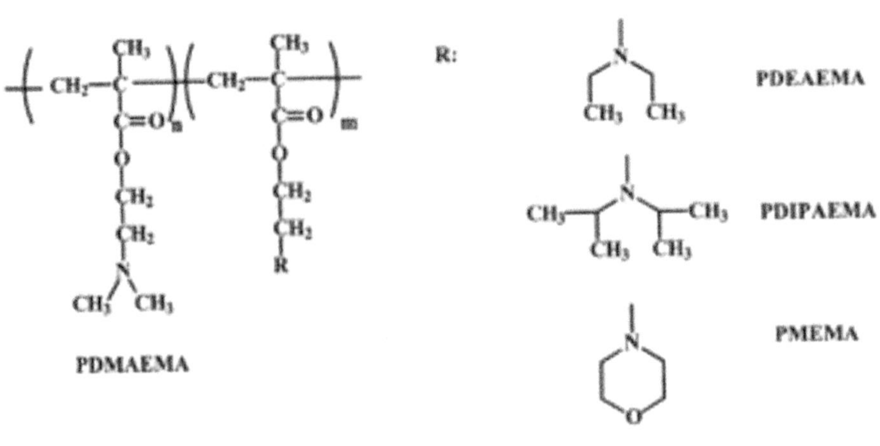

Scheme 33

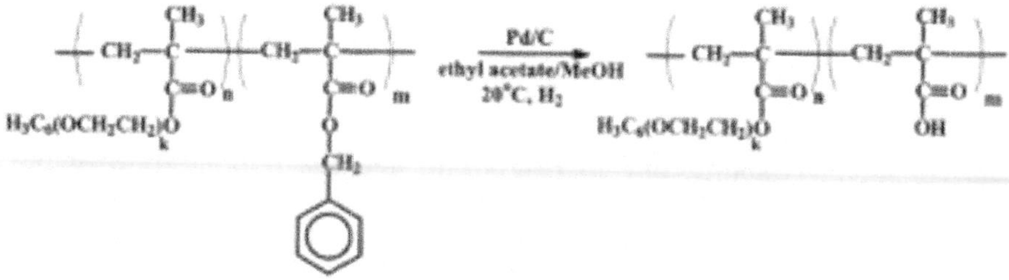

Scheme 34

and catalyst system. Well-defined terpolymers with predicted molecular weights and narrow molecular weight distributions were obtained in all cases. However, the samples possessed low molecular weights, due to the limitations of the method.

2.6 Synthesis of Block Copolymers by Olefin Metathesis Polymerization

The continuous developments in the field of metal-mediated olefin metathesis added novel tools to the arsenal of synthetic polymer chemistry. The vast body of research has focused on the ring opening metathesis polymerization, ROMP, of cyclic strained olefins [90–92]. When these olefins are employed ring cleavage leads to the formation of a difunctional moiety which effectively forms the building block of a polymer chain, as illustrated in Scheme 35. The strain release upon polymerization is an additional driving force shifting the reaction equilibrium in favor of the polymer. However, for monomers having low ring strain, such as cyclopentene, this process becomes reversible, resulting in monomer-polymer equilibrium. Several complexes of molybdenum, tungsten, titanium etc., have been employed as metathesis catalysts. In many cases this polymerization method is accompanied by a series of side reactions, (e.g. chain transfer, chain scission, ring-chain equilibrium, β-hydrogen transfer from the carbine to the metal) followed by olefin elimination, cyclopropanation, involving reaction of the carbine-metal species with the olefinic substrate etc. Consequently, this polymerization method was not able to promote the synthesis of well-defined polymers. However, recent advances in the synthesis of new catalytic systems, especially the Schrock- and Grubbs-

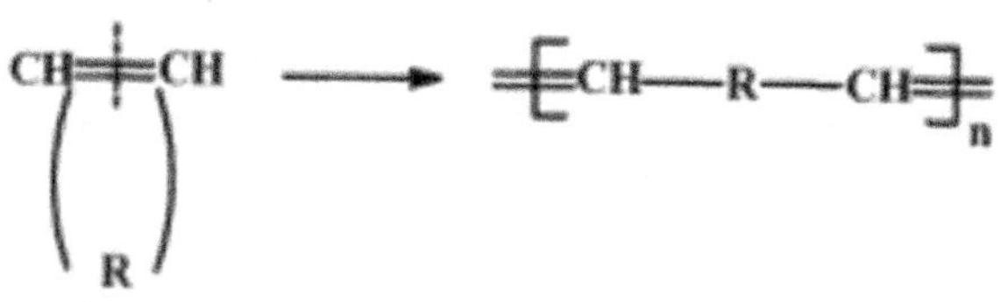

Scheme 35

types of catalysts have provided fascinating opportunities in the preparation of novel block copolymers [93, 94]. These "living" or better "controlled" polymerization systems have been successfully applied for the polymerization of norbornene and their derivatives. This behavior is attributed to the enhanced polymerizability, due to ring strain and the minimization of the side reactions, inhibited by increased steric hindrance due to branching at the α-carbons. An alternative and more recent approach employing olefin metathesis is the acyclic diene metathesis, ADMET [95–97], an analogous polycondensation reaction of α,ω-dienes utilizing an elimination reaction (Scheme 36). Since this procedure is an equilibrium process the olefinic by-product has to be selectively removed to drive the reaction towards the synthesis of high molecular weight products. As a polycondensation reaction it usually leads to chains having broad molecular weight distributions. In combination with other polymerization techniques it may lead to the synthesis of block copolymers.

The commercially available ROMP initiator Mo(NAr)($CHCMe_2Ph$)O-*t*-$Bu)_2$ (Ar=2,6-diisopropylphenyl-) was employed to polymerize cyclopentene at room temperature in the presence of PMe_3, which is known to bind reversibly to the propagating species [98]. The catalyst sites bound by PMe_3 are less active and slow down the propagation relative to the initiation reaction. Consequently, better control is achieved leading to polymers with low polydispersities. A major problem encountered for the polymerization of cyclopentene is the existence of the equilibrium between polymerization and depolymerization reactions. Thus, cyclopentene polymerizations must be conducted above the equilibrium monomer concentrations and the conversions well below the equilibrium value to avoid broadening of the molecular weight distribution through the depolymerization procedure. Therefore, the conversions were limited to low values (up to 40%). For this reason the synthesis of poly(ethylidenenorbornene)-*b*-polycyclopentene, PEN-*b*-PCP, had to begin with the polymerization of the EN monomer (Scheme 37), which was quantitatively polymerized. Narrow molecular weight distributions were obtained in all cases. Small high molecular weight peaks were observed, due to coupling of two polymer chains through bimolecular termination with trace oxygen. The polymers were quenched with propionaldehyde to end-cap the PCP blocks with propyl groups. Upon hydrogenation with Pd/$CaCO_3$ catalyst the PCP blocks were transformed to linear polyethylene.

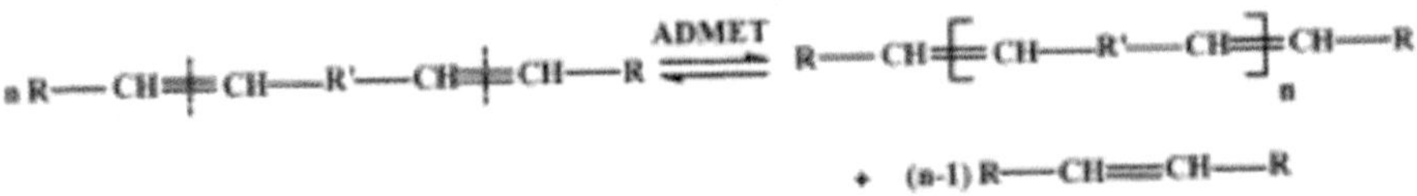

Scheme 36

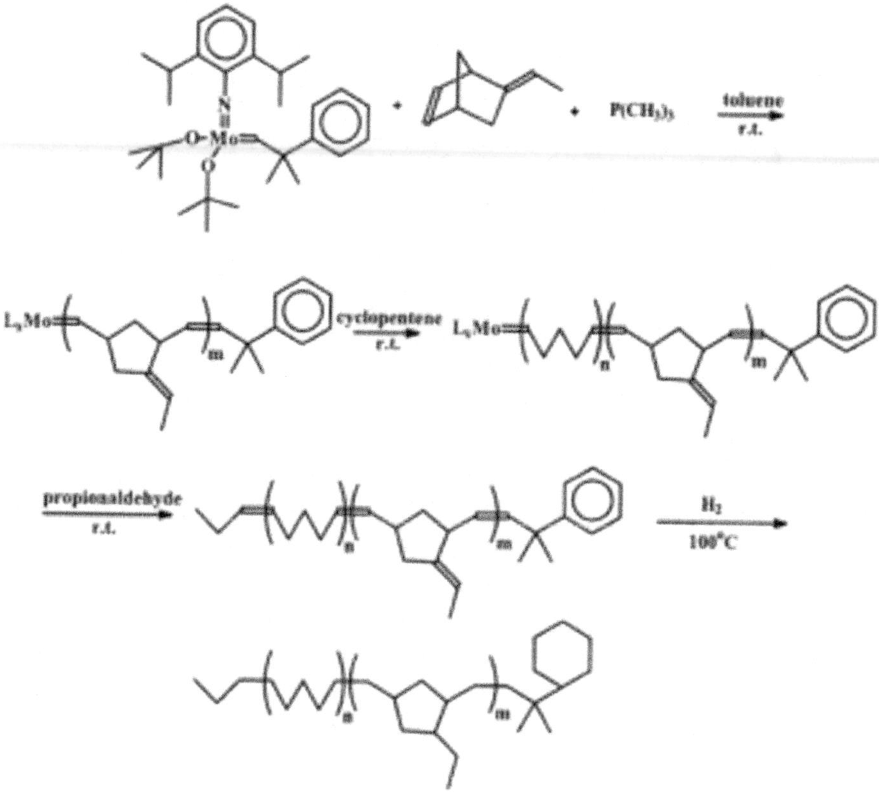

Scheme 37

Block copolymers produced from the sequential polymerization of *exo*-*N*-butyl-7-oxabicyclo[2.2.1]hept-5-ene-2,3-dicarboximide and another norbornene derivative carrying an adenine group were employed using a ruthenium catalyst, as shown in Scheme 38 [99]. The first monomer was polymerized in CH_2Cl_2 at room temperature followed by the addition of the second monomer and succinimide. In the presence of succinimide the conversion of the adenine-containing monomer was almost quantitative, probably due to the formation of hydrogen bonds with the adenine moiety, which prevents any undesired interaction with the catalyst. Monomodal peaks were obtained by SEC analysis with polydispersity indices ranging from 1.20 up to 1.60.

A carbazole-functionalized norbornene derivative, 5-(*N*-carbazoyl methylene)-2-norbornene, CbzNB, was polymerized via ROMP using the ruthenium catalyst $Cl_2Ru(CHPh)[P(C_6H_{11})_3]_2$ [100]. The polymerization was conducted in CH_2Cl_2 at room temperature, to afford products with polydispersity indices close to 1.3. Subsequent addition of 5-[(trimethylsiloxy)methylene]-2-norbornene showed a clear shift of the SEC trace of the initial polymer, indicating that a diblock copolymer was efficiently prepared in high yield.

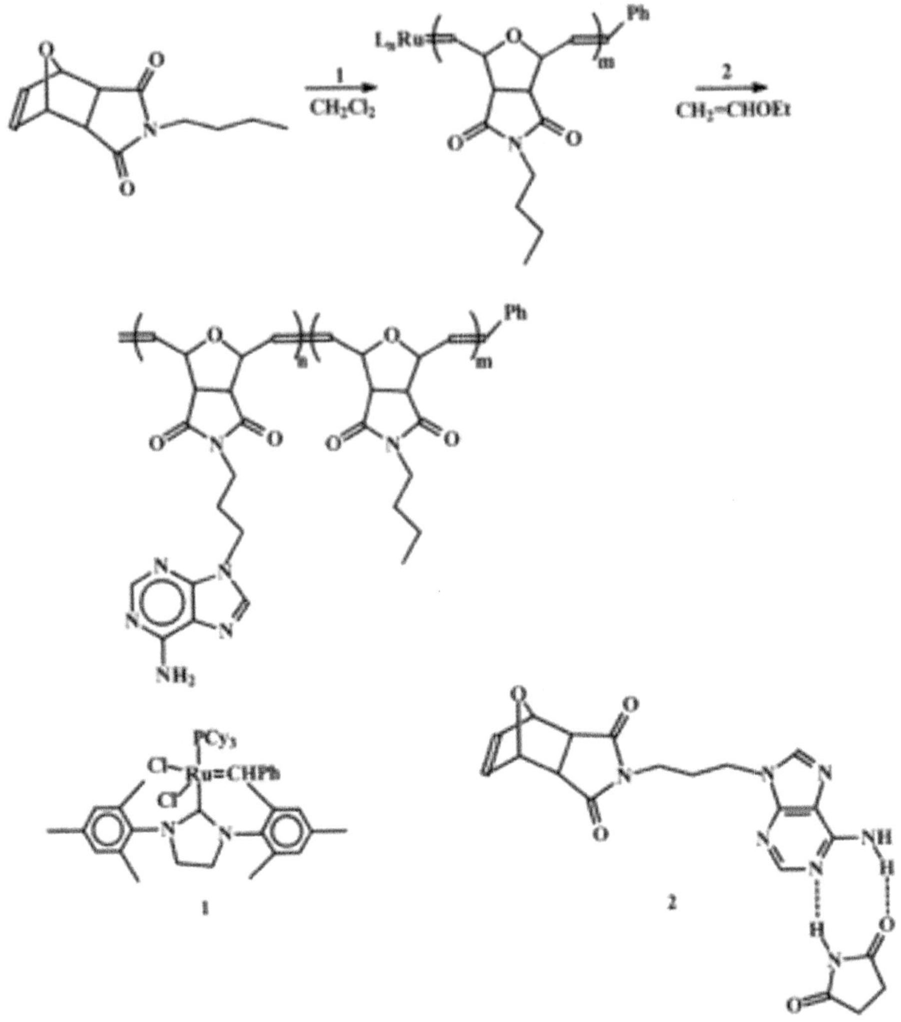

Scheme 38

Hydrolysis of the trimethylsilyl groups produced the corresponding hydroxyl groups. The reaction series is given in Scheme 39.

Synthesis of block copolymers of norbornene derivatives, with different side groups, has been reported via ROMP [101]. Initially, *exo-N*-butyl-7-oxabicyclo[2.2.1]hept-5-ene-2,3-dicarboximide was polymerized in acetone at room temperature with a ruthenium initiator (Scheme 40). The conversion of the reaction was quantitative. Subsequent addition of norbornene derivative carrying a ruthenium complex led to the formation of block copolymers in 85% yield. Due to the presence of ruthenium SEC experiments could not be performed. Therefore, it was not possible to determine the molecular weight

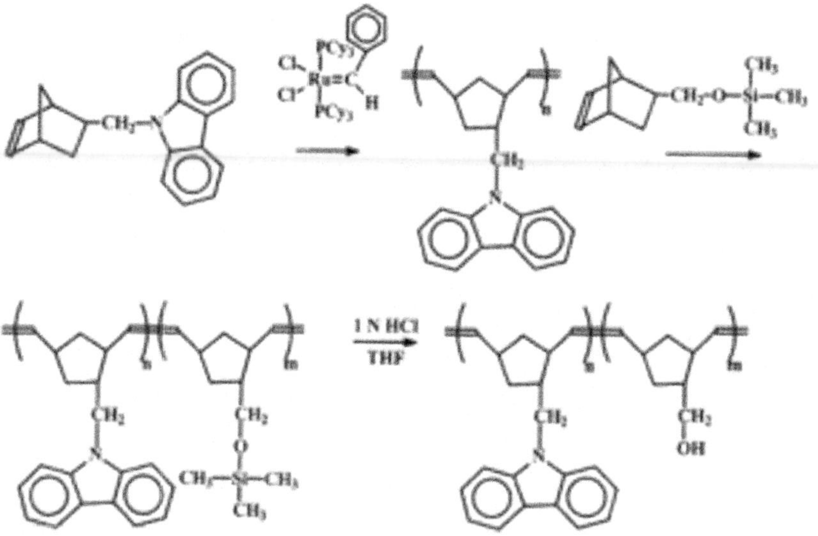

Scheme 39

distribution of the copolymers. End-group analysis by NMR gave evidence regarding the molecular weight of the samples ($M_n = 28\,000$).

The living character of the ROMP promoted by the initiator $Ru(CHPh)(Cl)_2(PCy_3)_2$ (Cy = cyclohexane) was tested with the synthesis of diblock, triblock, and tetrablock copolymers of norbornene derivatives carrying acetyl-protected glucose, [2,3,4,6-tetra-*O*-acetyl-glucos-1-*O*-yl 5-norbornene-2-carboxylate], A or maltose groups, [2,3,6,2′,3′,4′,6′-hepta-*O*-acetyl-maltos-1-*O*-yl 5-norbornene-2-carboxylate], B, shown in Scheme 41 [102]. The AB, ABA, and ABAB structures were prepared by sequential addition of monomers with narrow molecular weight distributions to quantitative conversions.

2.7 Synthesis of Block Copolymers by the Post-Polymerization Formation of Metal Complexes

The ability of a broad range of *N*-heterocycles to act as effective complexation agents for several transition metal ions has been known for many years. Such behavior was later used in supramolecular chemistry for the construction of complex architectures [103]. This knowledge has been transferred to polymer chemistry with the development of metal-complexing and metal-containing polymers. The main objective is to combine the polymer properties of the architecture formed with the reversible binding from the supramolecular entities connected to the polymer backbone. In supramolecular chemistry the linkages can be formed or broken by tuning the external stimuli. 2,2′-

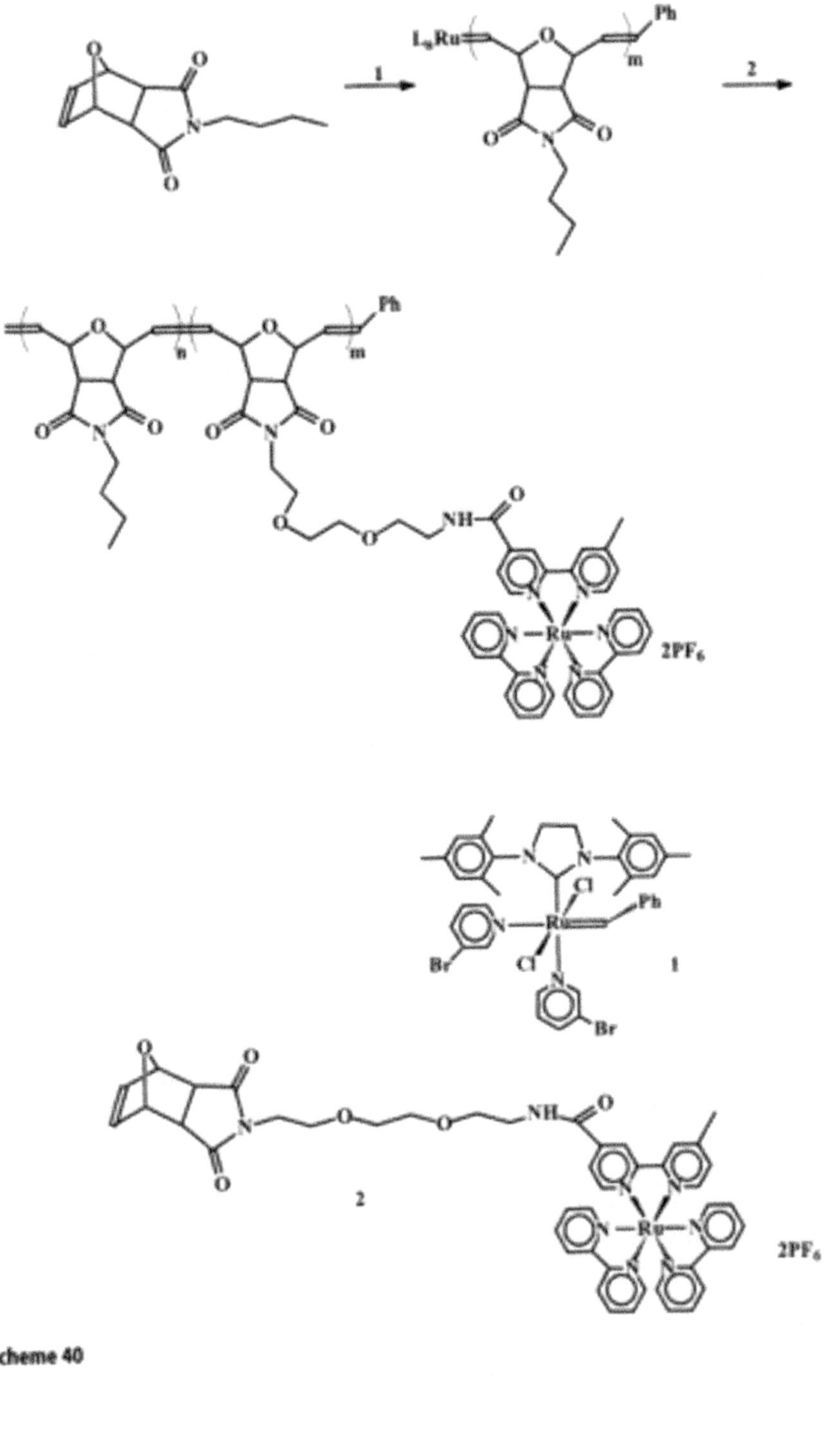

Scheme 40

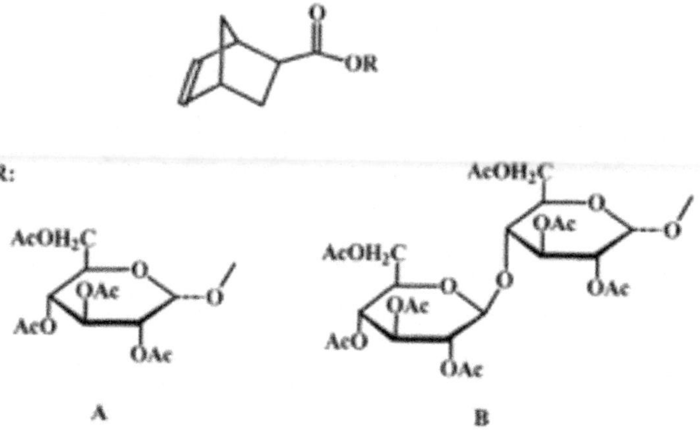

Scheme 41

Bipyridines were efficiently used in supramolecular chemistry [104]. Since the molecule is symmetric no directed coupling procedure is possible. In addition, 2,2′ : 6′,2″-terpyridine ligands can lead to several metal complexes, usually bis-complexes having octahedral coordination geometries [105, 106]. Lifetimes of the metal-polymeric ligand depend to a great extent on the metal ion used. Highly labile complexes as well as inert metal complexes have been reported. The latter case is very important since the complexes can be treated as conventional polymers, while the supramolecular interaction remains present as a dormant switch.

A general strategy developed for the synthesis of supramolecular block copolymers involves the preparation of macromolecular chains end-capped with a 2,2′ : 6′,2″-terpyridine ligand which can be selectively complexed with $RuCl_3$. Under these conditions only the mono-complex between the terpyridine group and Ru(III) is formed. Subsequent reaction with another 2,2′ : 6′,2″-terpyridine terminated polymer under reductive conditions for the transformation of Ru(III) to Ru(II) leads to the formation of supramolecular block copolymers. Using this methodology the copolymer with PEO and PS blocks was prepared (Scheme 42) [107].

The synthesis of the supramolecular block copolymer PEO-[Ru]-poly-(ethylene-co-butylene) was described employing the same procedure [108].

Using a diblock copolymer PS-*b*-P2VP, instead of PS, the supramolecular triblock terpolymer PEO-[Ru]-P2VP-*b*-PS was prepared, as reported in Scheme 43 [109]. The α-methoxy-ω-(2,2′ : 6′,2″-terpyridinyl)oxy-PEO was obtained as previously reported. Living anionic polymerization was utilized for the synthesis of hydroxy-terminated PS-*b*-P2VP chains. *t*BuOK was used to give the corresponding alkoxy end-group followed by reaction with 4′-chloroterpyridine to yield the terpyridine-terminated diblock. Combination of the two end-functionalized polymers under reductive conditions provided the desired supramolecular structure PEO-[Ru]-P2VP-*b*-PS.

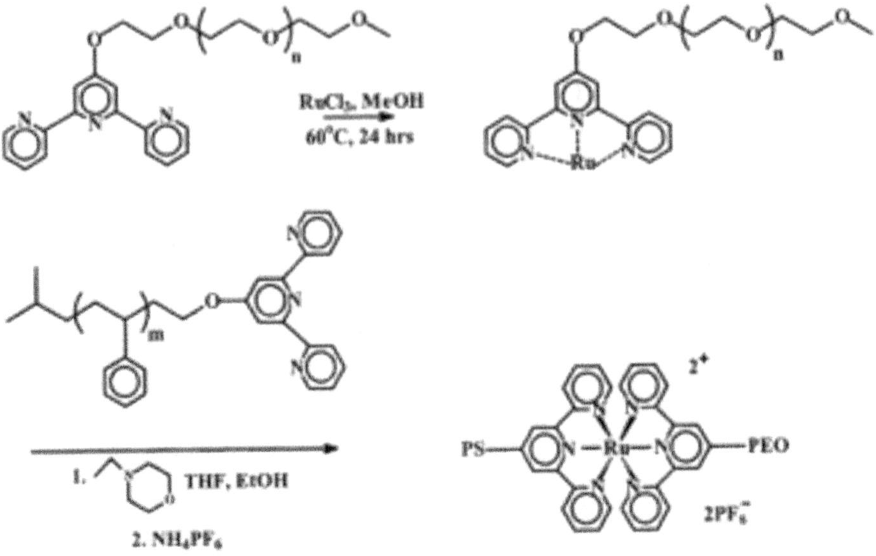

Scheme 42

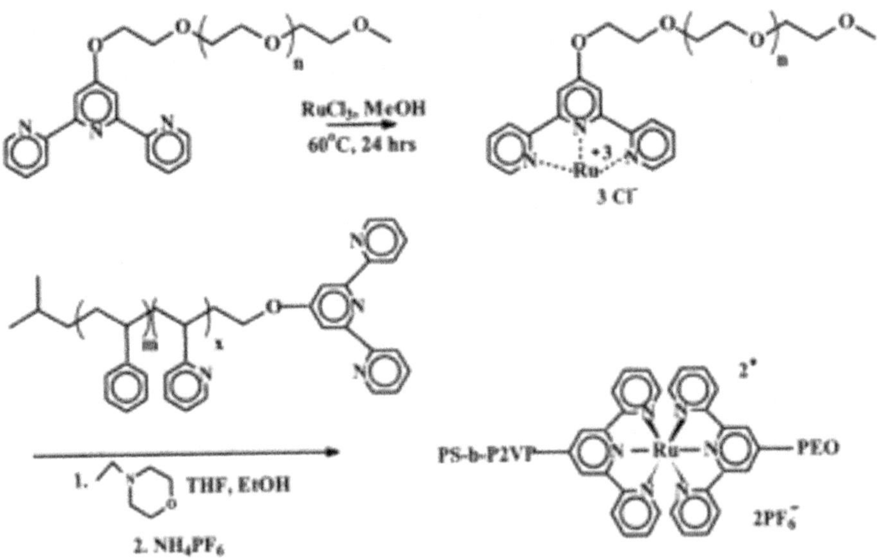

Scheme 43

2.8 Synthesis of Block Copolymers by Transition Metal-Catalyzed Polymerization

Transition metal polymerization catalysts have stimulated tremendous efforts in academic research resulting in numerous industrial applications. Ziegler-Natta and metallocene catalysts have been used for the synthesis of tailor-made polymers regarding the microstructure, comonomer incorporation and composition, end-group functionality and molecular weight. Coordination polymerization has been expanded to a wide variety of monomers, such as olefins, styrenes, dienes, (meth)acrylates, lactones, lactames, carbonates etc. Newer advances in the synthesis of novel catalytic systems and the study of the polymerization mechanism have allowed for the realization of controlled or even "living" polymerization leading to the synthesis of more complex structures, such as block and graft copolymers. Despite the fact that these systems, compared to classic living polymerizations, lack the high degree of control regarding the molecular characteristics, the chemical and compositional homogeneity, recent landmark discoveries in the field, offer attractive potential for the development of versatile polymerization systems expanding the frontiers of polymer science.

The polymerization of 1,5-hexadiene with the titanium catalyst, shown in Scheme 44 in the presence of methylaluminoxane, MAO, at 0 °C was investigated by NMR and SEC techniques [110]. The structural analysis showed that the polymer, obtained under these conditions, was comprised of methylene-1,3-cyclopentane, MCP (63%) and 3-vinyl tetramethylene, VTM (37%) units. Samples with high molecular weights and relatively narrow molecular weight distributions (lower than 1.3) were obtained. The controlled character promoted by this catalytic system was exploited by the sequential polymerization of propylene and 1,5-hexadiene to form block copolymers consisting of a *syndio*-polypropylene block and another one with MCP and VTM units. Copolymers of narrow molecular weight distributions were prepared with this method. However, there was no report concerning the yield of the copolymerization and the correspondence between the stoichiometric and the experimentally observed molecular weights.

An allyl samarocene catalyst, $[(CMe_2C_5H_4)_2SmCl(C_3H_5)MgCl_2(THF)_4$, was employed for the synthesis of *trans*-PI-*b*-PCL copolymers and poly(*trans*-isoprene-*co*-hex-1-ene)-*b*-PCL terpolymers [111]. The copolymerizations

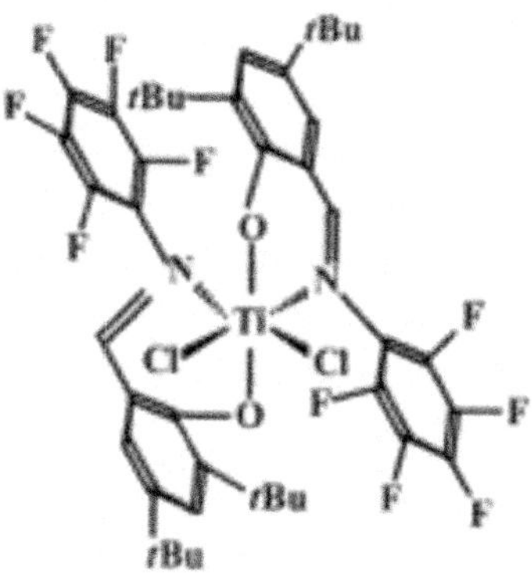

Scheme 44

were conducted in toluene, or THF at 50 °C by the sequential addition of monomers, starting from the polymerization of isoprene or isoprene/hex-1-ene. The inverse addition does not induce block copolymerization and leads only to the formation of PCL homopolymer. Molecular weights could not be easily controlled since the yields of the copolymerization were rather low in most cases (less than 50%). Furthermore, broad molecular weight distributions, between 1.5 and 2.0 were obtained. The final products were used as compatibilizers for PCL and PI blends.

The zirconocene catalyst $Me_2C(Cp)(Ind)ZrMe_2$ was employed for the synthesis of PE-*b*-PMMA block copolymers in the presence of $B(C_6F_5)_3$ by sequential addition of monomers [112]. Sampling from the reactor, before the addition of MMA, was not performed to enable the chromatographic analysis of the PE block. No homopolymer was traced after the extraction with selective solvents. However, the possibility of the formation of a random sequence of the ethylene and MMA units, due to incomplete polymerization of the first monomer prior to the addition of the second, cannot be ruled out. Copolymers having relatively low molecular weights and broad molecular weight distributions were obtained ($M_w/M_n > 2.4$).

Block copolymers P*n*BuMA-*b*-PMMA were prepared using zirconocene catalysts [113]. In a previous study using the $Cp_2ZrMe_2/B(C_6F_5)_3$ system as catalyst/co-catalyst and CH_2Cl_2 as the solvent, the synthesis of the PMMA-*b*-P*n*BuMA block copolymers starting from MMA polymerization was not successful [114]. The reverse order of addition was thus employed with the catalytic systems $Cp_2ZrMe_2/B(C_6F_5)_3/ZnEt_2$ (1), *rac*-$Et(Ind)_2ZrMe_2/B(C_6F_5)_3/ZnEt_2$ (2) and *rac*-$Et(Ind)_2ZrMe_2/[Me_2NHPh]^+[B(C_6F_5)_4]^-/ZnEt_2$ (3). $ZnEt_2$ acts both as an activator of the methacrylate monomer and as an internal scavenger, reacting with the impurities. Previous kinetic experiments have indicated the exact experimental conditions (temperature, nature of catalytic system, monomer, catalyst and co-catalyst concentration etc.), under which the polymerization was completed, and the side reactions were prevented. Catalytic system (1) led to polydispersities of approximately 1.3, whereas catalytic systems (2) and (3) promoted, on one hand, the isotactic polymerization of methacrylates and on the other the better polymerization control, leading to very narrow molecular weight distributions ($M_w/M_n = 1.10$). SEC analysis and DSC measurements revealed that well-defined structures were indeed prepared.

The samarocene complexes $SmMe(C_5Me_5)_2(THF)$ and $[SmH(C_5Me_5)_2]_2$ were employed as initiators for the synthesis of the well-defined block and triblock copolymers poly(trimethylsilyl methacrylate)-*b*-PMMA, PTMSMA-*b*-PMMA, PMMA-*b*-PTMSMA, PTMSMA-*b*-P*n*BuA, PMMA-*b*-P*n*BuA, and PMMA-*b*-P*n*BuA-*b*-PMMA [115]. When the procedure started with MMA polymerization, followed by the addition of TMSMA, block copolymers with low polydispersity and stoichiometric molecular weights in very good agreement with the experimental values were obtained. The reverse mode of addition leads to broader distributions and molecular weights higher than the stoichiometric values. In all cases the polymers were highly syndiotactic.

2.9 Synthesis of Block Copolymers by Combinations of Different Polymerization Techniques

Every polymerization method is limited to a certain type and number of monomers, thus preventing the possibility to synthesize block copolymers with a wide combination of monomers. However, recent advances in polymer synthesis enabled the switching of the polymerization mechanism from one type to another, thereby permitting the preparation of block copolymers composed of monomers that can be polymerized by different techniques.

The transformation of the chain end active center from one type to another is usually achieved through the successful and efficient end-functionalization reaction of the polymer chain. This end-functionalized polymer can be considered as a macroinitiator capable of initiating the polymerization of another monomer by a different synthetic method. Using a semitelechelic macroinitiator an AB block copolymer is obtained, while with a telechelic macroinitiator an ABA triblock copolymer is provided. The key step of this methodology relies on the success of the transformation reaction. The functionalization process must be 100% efficient, since the presence of unfunctionalized chains leads to a mixture of the desired block copolymer and the unfunctionalized homopolymer. In such a case, control over the molecular characteristics cannot be obtained and an additional purification step is needed.

Poly(ethylene glycol)-*b*-poly(2-methyl-2-oxazoline), PEG-*b*-PMeOx block copolymers were synthesized through the anionic ring opening polymerization of ethylene oxide using potassium 3,3-diethoxypropanolate. Subsequent reaction with methanesulfonyl chloride provided the heterotelechelic acetal-PEG-SO_2CH_3 polymer. NMR analysis revealed that this procedure is efficient and leads to quantitative functionalization. This product was then used as the macroinitiator for the cationic ring opening polymerization of MeOx in nitromethane at 60 °C [116]. SEC analysis proved the procedure efficient, without termination reactions. Low molecular weight copolymers with polydispersities around 1.4 were obtained with this procedure. Alkaline hydrolysis of the amide group of the PMeOz block to a secondary amino group resulted in the synthesis of poly(ethylene oxide-*b*-ethylenimine) block copolymers (Scheme 45).

Anionically prepared hydroxy-terminated PBd was reacted with $AlEt_3$ to form the corresponding aluminum alkoxide macroinitiator, capable of initiating the polymerization of L-lactide [117]. Using ratios $[PBd-OH]/[AlEt_3]$ between 1 and 6, reaction temperatures between 70 and 120 °C and maintaining the conversion of the lactide polymerization below 90%, products with narrow molecular weight distribution were obtained.

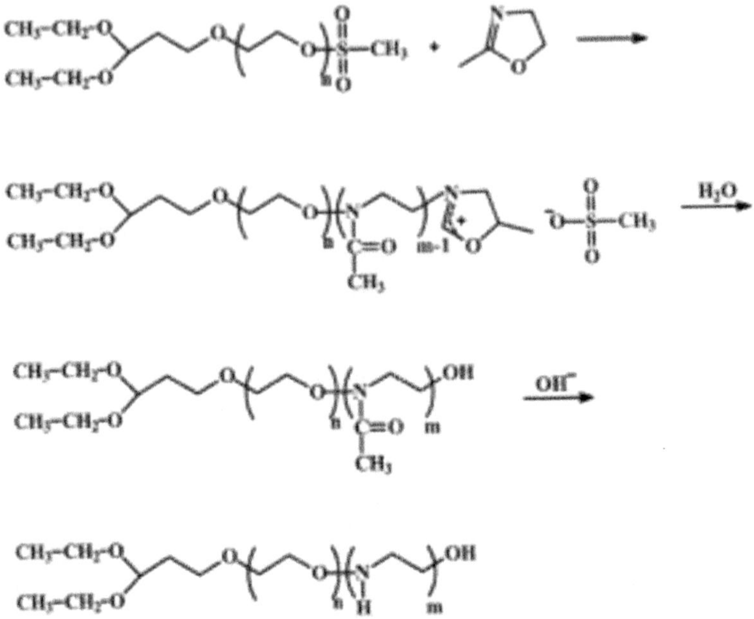

Scheme 45

Transformation of living anionic polymerization into ROMP was employed for the synthesis of block copolymers consisting of PEO and polynorbornene derivative blocks. EO was polymerized using diphenylmethylpotassium as the initiator and was terminated by vinylbenzyl chloride to produce a PEO macromonomer with a terminal vinyl group. This macromonomer was transformed into a ROMP macroinitiator by an alkylidene exchange reaction between the propylidene complex $RuCl_2(=CHC_2H_5)(PCy_3)_2$ and the PEO macromonomer. The macroinitiator was subsequently used for the polymerization of norbornene derivatives to give the desired block copolymers, as illustrated in Scheme 46. SEC analysis showed that in a few cases only a small amount (1–2%) of PEO homopolymer was present, probably due to incomplete functionalization at the initial step.

1,2-bis(2′-Bromobutyryloxy)ethane was used as a difunctional initiator for the ATRP of styrene to give polymers terminated with bromine groups at both ends [119]. These functions were then reacted with silver perchlorate giving a macromolecular initiator suitable for the cationic ring opening polymerization of THF leading to PTHF-*b*-PS-*b*-PTHF triblock copolymers (Scheme 47). SEC analysis indicated the presence of single peaks, but with moderate polydispersities (between 1.3 to 1.4).

A combination of ATRP and ROP was employed for the synthesis of PLLA-*b*-PS block copolymers and PLLA-*b*-PS-*b*-PMMA triblock terpolymers [120]. Styrene was initially polymerized using the functional initiator β-hydroxyethyl α-bromobutyrate, HEBB, and the catalytic system CuBr/bpy.

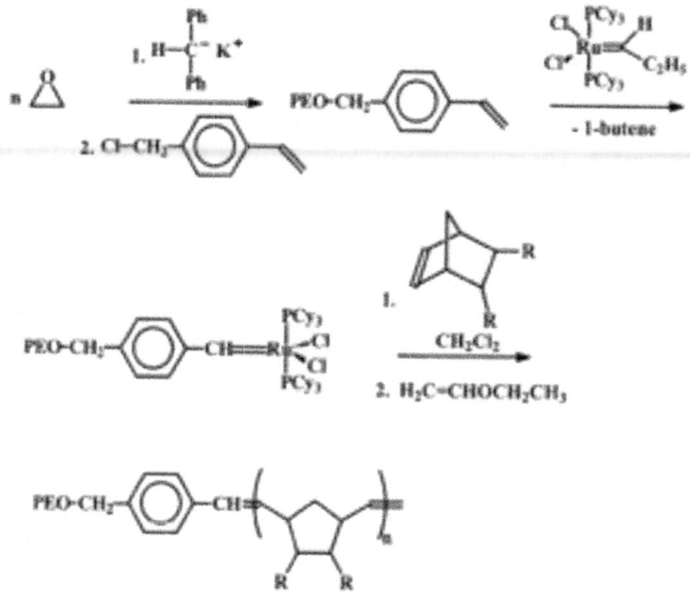

Scheme 46

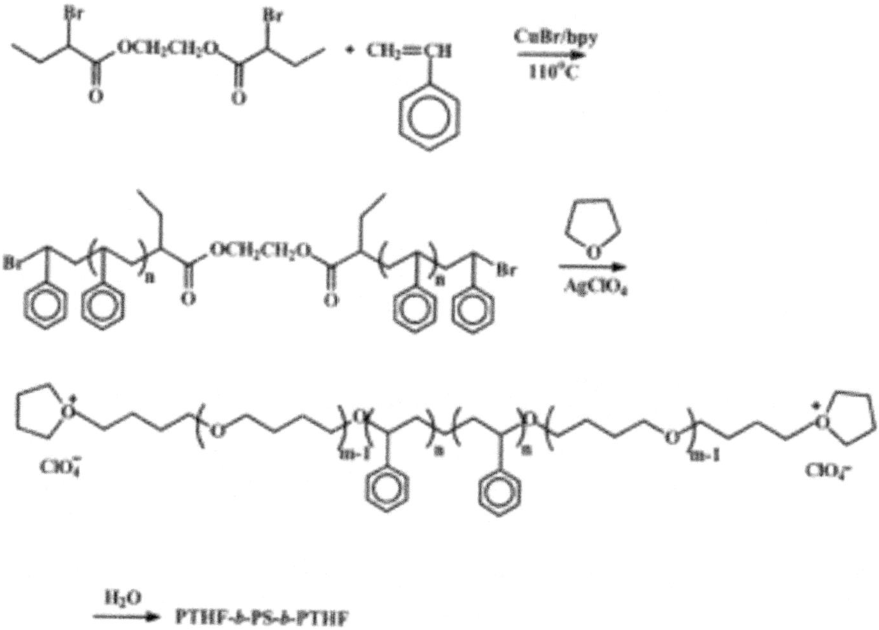

Scheme 47

Under these conditions the living polymerization of styrene was promoted leading to polymers with end-hydroxyl groups. These products subsequently served as macroinitiators for the polymerization of L-lactide in the presence of $Sn(OCt)_2$ as the catalyst at 115 °C, leading to the formation of PLLA-*b*-PS block copolymers. The PS block carrying a hydroxyl group at the one end and a bromine group at the other end can be used for the polymerization of MMA by ATRP before the polymerization of L-lactide to produce PLLA-*b*-PS-*b*-PMMA triblock terpolymers (Scheme 48). In all cases products with narrow molecular weight distributions were obtained.

Free radical polymerization combined with anionic ring polymerization was employed for the synthesis of poly(*N*-vinylpyrrolidone)-*b*-poly(D,L-lactide), PVP-*b*-PDLLA, as shown in Scheme 49 [121]. The free radical polymerization of VP was conducted using 2,2′-azobis[2-methyl-*N*-(2-hydroxyethyl)propionamide] as the initiator, isopropyl alcohol and 2-

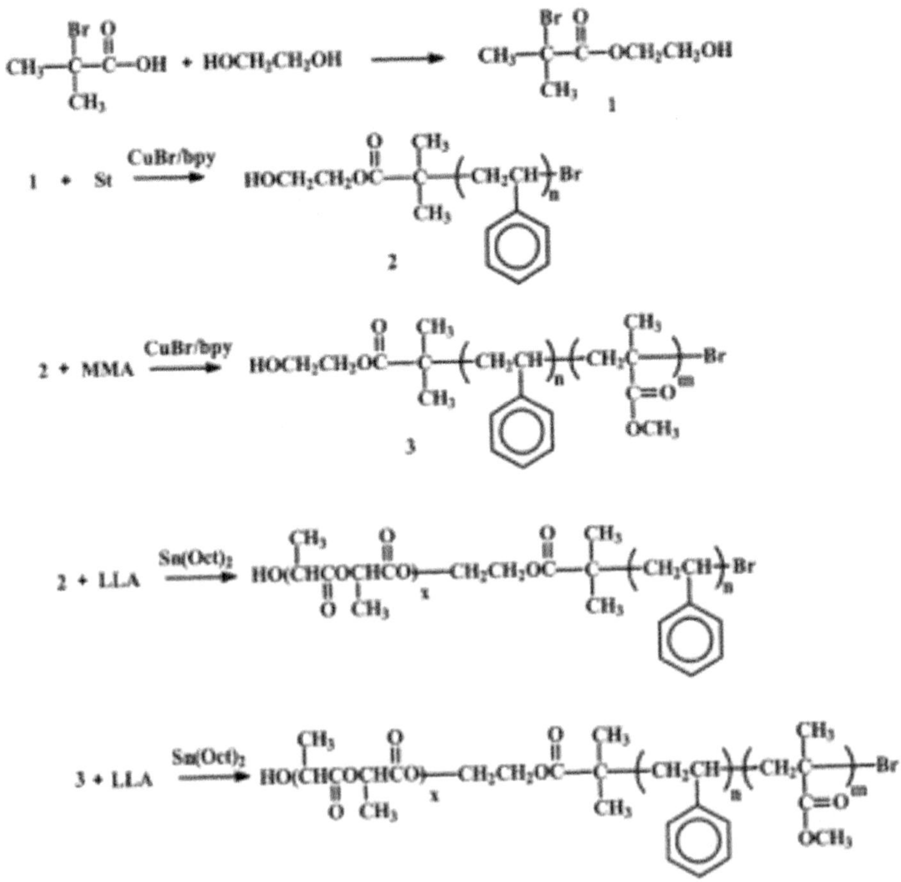

Scheme 48

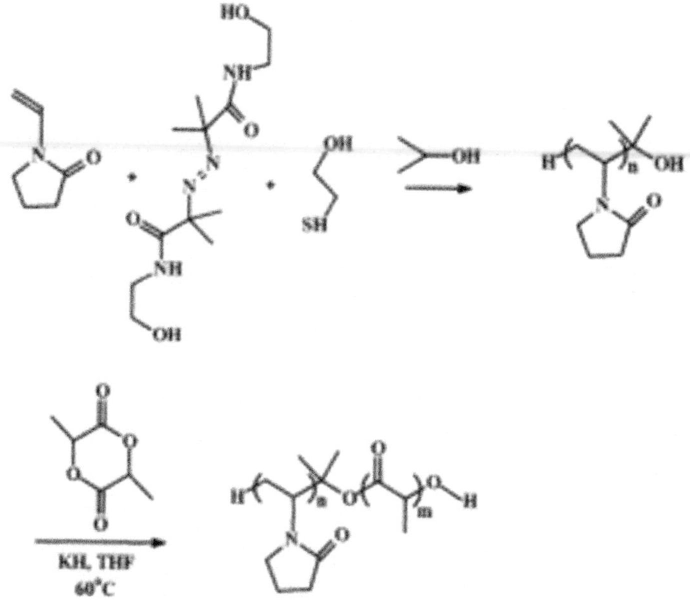

Scheme 49

mercaptoethanol as the chain transfer agents to provide polymers with the hydroxyl end group. This end function was activated with KH in order to initiate the anionic ring opening polymerization of DLLA. As expected broad molecular weight distributions were obtained for the first block (M_w/M_n higher than 1.5), while the distributions for the final products were narrower ($1.14 < M_w/M_n < 1.48$). The broad distribution of the first block indicates the increased compositional heterogeneities of the samples.

ABA triblock copolymers, where A was PBd and B either PS or PMMA were prepared by the combination of ROMP and ATRP techniques [122]. The PBd middle blocks were obtained through the ROMP of cyclooctadiene in the presence of 1,4-chloro-2-butene or cis-2-butene-1,4-diol bis(2-bromo)propionate using a Ru complex as the catalyst. The end allyl chloride or 2-bromopropionyl ester groups were subsequently used for the ATRP of either styrene or MMA using CuX/bpy (X = Cl or Br) as the catalytic system (Scheme 50). Quantitative yields but rather broad molecular weight distributions (M_w/M_n higher than 1.4) were obtained.

Metallocene catalysis has been combined with ATRP for the synthesis of PE-*b*-PMMA block copolymers [123]. PE end-functionalized with a primary hydroxyl group was prepared through the polymerization of ethylene in the presence of allyl alcohol and triethylaluminum using a zirconocene/MAO catalytic system. It has been proven that with this procedure the hydroxyl group can be selectively introduced into the PE chain end, due to the chain transfer by $AlEt_3$, which occurs predominantly at the dormant end-

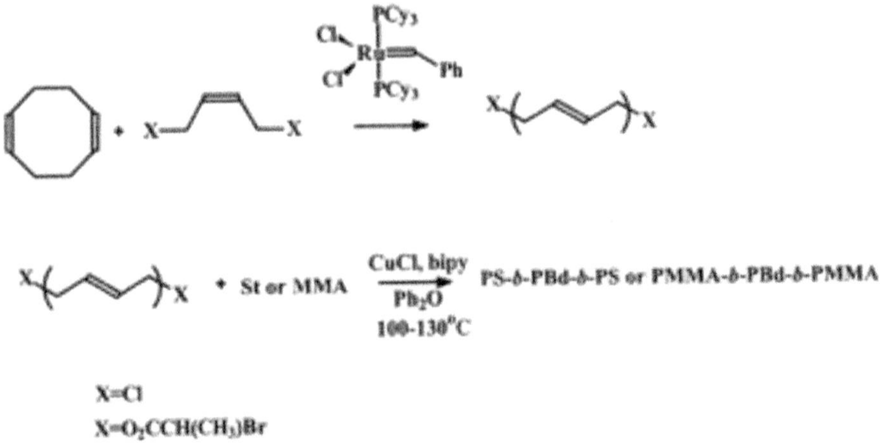

Scheme 50

incorporated masked allyl alcohol. This terminal hydroxyl group was subsequently reacted with 2-bromoisobutyryl bromide to form the macroinitiator, capable of initiating the ATRP of MMA in the presence of either CuBr/PMDETA or $RuCl_2(PPh_3)_3/Bu_2NH$ as the catalytic system. The reaction series is given in Scheme 51. The polymerization was performed at 120 °C with *o*-xylene as the solvent. Higher conversions were obtained with the Ru catalytic system. However, in most cases the conversions were relatively low. In addition very broad molecular weight distributions, higher than 2.0, were obtained in all cases. These structures were examined as compatibilizers for PE and PMMA blends.

In a series of papers poly(ethylene oxide), PEO, or poly(ethylene glycol), PEG and poly(propylene oxide) having one or two terminal hydroxyl groups were used as macroinitiators for the synthesis of diblock or triblock copolymers by ATRP after transformation of the end functions to bromide or chloride groups, which then initiated the polymerization of other monomers. Reaction of semitelechelic PEG with hydroxyl end groups with 2-bromoisobutyryl bromide yielded a macroinitiator, which was able to polymerize several methacrylates, such as MMA, EMA, EA, *t*BuMA, and DMAEMA [124]. The copolymerizations were conducted either in bulk or in solution using PMDETA as the ligand and CuBr as the catalyst at 60 °C. The copolymerization reaction was left to run overnight. The following structures were finally obtained: PEG-*b*-PMMA, PEG-*b*-PEMA, PEG-*b*-PEA, PEG-*b*-P*t*BuMA, PEG-*b*-PDMAEMA, PEG-*b*-PMMA, PEG-*b*-(PMMA-*co*-*t*BuMA), PEG-*b*-(PEA-*co*-*t*BuMA), and PEG-*b*-(PDMAEMA-*co*-EMA). The general reaction sequence is reported in Scheme 52. The synthesis of the macroinitiator was successful and the copolymerization conversions near quantitative, except in the case of DMAEMA and EMA which showed rather low conversions (55–70%). Rather high polydispersities, between 1.3 and 1.6, were obtained,

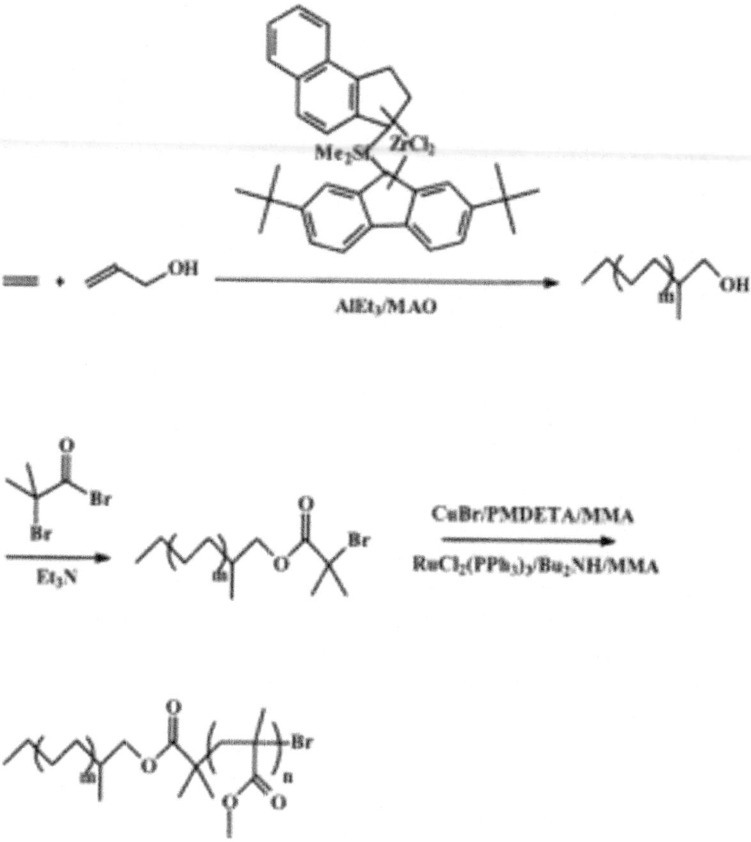

Scheme 51

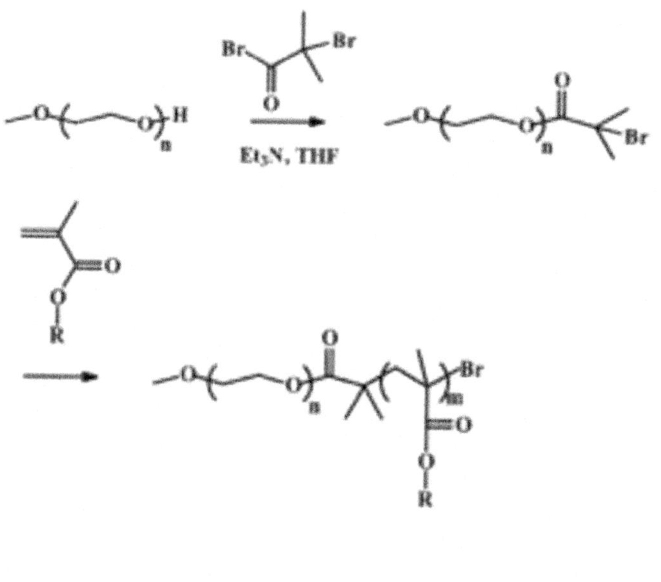

Scheme 52

and the molecular weights prepared were consistently lower than 10 000. The *t*-butyl groups of the *t*BuMA units were subsequently hydrolyzed under mild acidic conditions, and the dimethylamine groups of the DMAEMA units were quaternized using methyl iodide to give amphiphilic block copolymers with either anionic or cationic side groups. The micellar properties of these materials were examined in aqueous solutions.

Diblock copolymers PEO-*b*-PS have been prepared using PEO macroinitiator and ATRP techniques [125]. The macroinitiator was synthesized by the reaction of monohydroxy-functionalized PEO with 2-chloro-2-phenylacetylchloride. MALDI-TOF revealed the successful synthesis of the macroinitiators. The ATRP of styrene was conducted in bulk at 130 °C with CuCl as the catalyst and 2,2′ bipyridine, bipy, as the ligand. Yields higher than 80% and rather narrow molecular weight distributions ($M_w/M_n < 1.3$) were obtained. The surface morphology of these samples was investigated by atomic force microscopy, AFM.

A wide range of sugar-based block copolymers have been prepared using macroinitiators based on PEO, PPO, and PCL and even by sequential monomer addition of other methacrylic monomers, such as DMAEMA, 2-(diisopropylamino)ethyl methacrylate, DIPAEMA or glycerol monomethacrylate, GMA. 2-Gluconamidoethyl methacrylate, GAMA and 2-lactobionamidoethyl methacrylate, LAMA (Scheme 53) were used for the synthesis of diblocks and triblocks such as: PEO-*b*-PGAMA, PPO-*b*-PGAMA, PPO-*b*-PLAMA, PLAMA-*b*-PGAMA, PEO-*b*-PGAMA-*b*-PDMAEMA, PEO-*b*-PGAMA-*b*-PDIPAEMA, PEO-*b*-PGAMA-*b*-PGMA, PEO-*b*-PLAMA-*b*-PDMAEMA, PEO-*b*-PLAMA-*b*-PGAMA, and PGAMA-*b*-PCL-*b*-PGAMA [126]. The monofunctional macroinitiators of PEO and PPO and the difunctional macroinitiator of PCL were prepared by esterification of the corresponding end hydroxyl

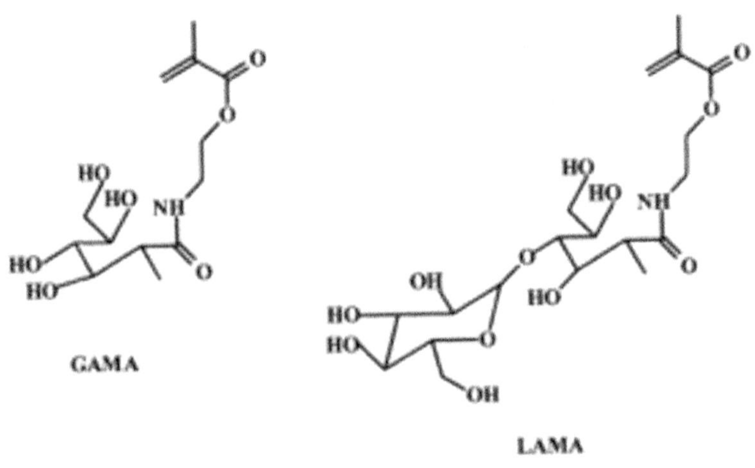

Scheme 53

groups with 2-bromoisobutyryl bromide. The block copolymerization reactions were conducted in methanol, water/methanol mixtures or *N*-methyl-2-pyrrolidone at 20 °C using CuBr as the catalyst and bpy as the ligand. Yields lower than 80% were obtained, and termination reactions were also observed during the synthesis. Relatively narrow molecular weight distributions were reported. The reversible micellar behavior of these samples was studied in aqueous solutions.

PPO-*b*-PDEAEMA block copolymers were prepared using a PPO macroinitiator, synthesized as previously described. The copolymerization was performed in methanol at 55 °C using CuCl as the catalyst and HMTETA as the ligand [127]. The yield was quantitative and the molecular weight distribution equal to 1.20.

The triblock terpolymer poly(propylene oxide)-*b*-poly[2-(dimethylamino)ethyl methacrylate]-*b*-poly[oligo(ethylene glycol) methacrylate], PPO-*b*-PDMAEMA-*b*-POEGMA, was prepared using the PPO macroinitiator followed by the addition of CuCl, HMTETA, and DMAEMA for the polymerization of the second block and finally OEGMA for the synthesis of the final product (Scheme 54) [128].

Employing similar procedures, PPO-*b*-POEGMA block copolymers and POEGMA-*b*-PPO-*b*-POEGMA triblock copolymers were prepared from the corresponding PPO macroinitiators [129]. The polymerizations were performed in a isopropanol/water (70/30) mixture at 20 °C using CuCl and bpy. The methacrylate monomer was almost quantitatively polymerized, and the polydispersities were lower than 1.25 in most cases. Less than 5% PPO homopolymer contamination was detected by SEC analysis.

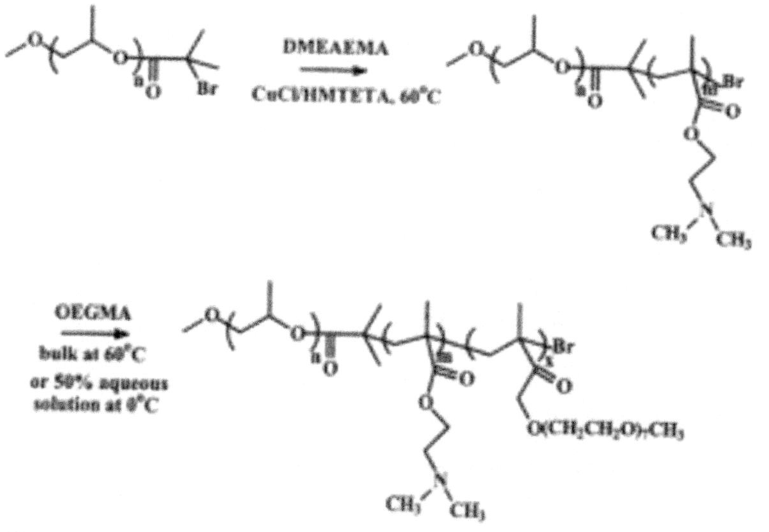

Scheme 54

A PEO macroinitiator with Si – H end groups was prepared through the condensation of monohydroxy-terminated PEO with ClSiMePhH in the presence of pyridine [130]. The presence of the Si – Ph moiety prevents the hydrolysis of the Si – O – C bond, due to steric factors. This macroinitiator was subsequently used for the synthesis of poly(ferrocenyldimethylsilane), PFS, to afford PEO-*b*-PFS block copolymers. The ROP of the ferrocenophane was conducted catalytically using the Pt(0) Karstedt's catalyst in toluene at 25 °C (Scheme 55). Rather broad molecular weight distributions (higher than 1.3) were obtained.

An interesting procedure has been proposed for the synthesis of amylose-*b*-PS block copolymers through the combination of anionic and enzymatic polymerization [131]. PS end-functionalized with primary amine or dimethylsilyl, – $SiMe_2H$ groups were prepared by anionic polymerization techniques, as shown in Scheme 56. The PS chains represented by the curved lines in Scheme 56 were further functionalized with maltoheptaose oligomer either through reductive amination (Scheme 57) or hydrosilylation reactions (Scheme 58). In the first case sodium cyanoborohydride was used to couple the saccharide moiety with the PS primary amine group.

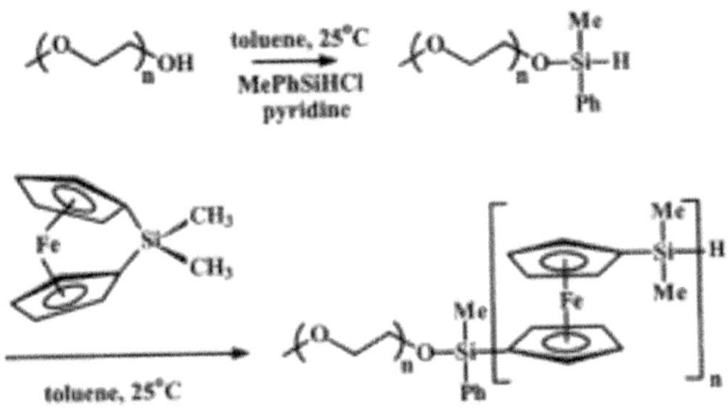

Scheme 55

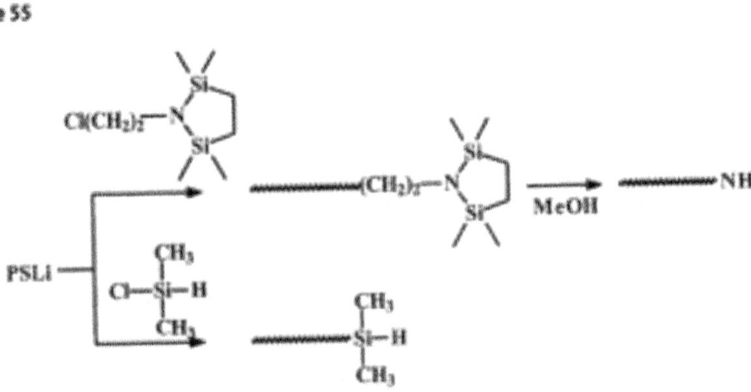

Scheme 56

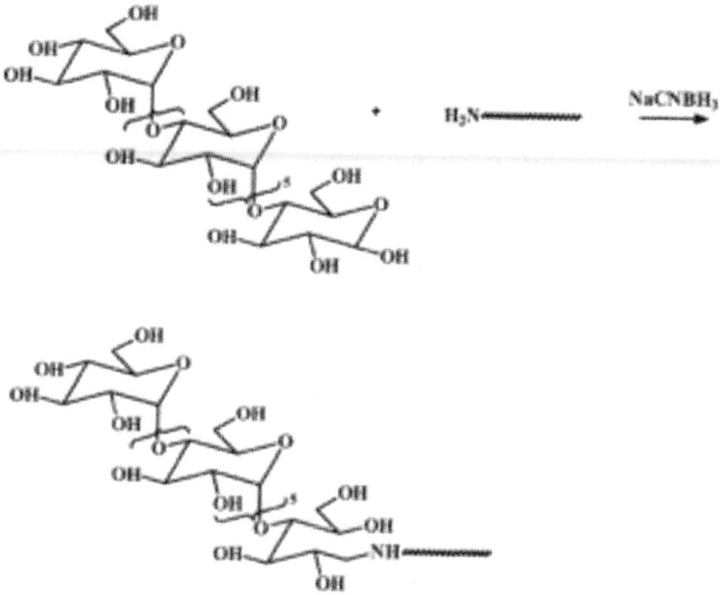

Scheme 57

For the hydrosilylation reaction various rhodium, platinum, and cobalt catalysts were employed. For the further chain extension the OH-functionalities were deprotected by KCN in methanol. The final step involved the enzymatic polymerization from the maltoheptaose-modified polystyrene using α-D-glucose-1-phosphate dipotassium salt dihydrate in a citrate buffer (pH = 6.2) and potato phosphorylase (Scheme 59). The characterization of the block copolymers was problematic in the case of high amylose contents, due to the insolubility of the copolymers in THF.

PIB-*b*-PEG block copolymers were prepared through the coupling reaction of end-functionalized PIB with $-SiMe_2H$ groups and allyl-terminated PEG via the hydrosilylation reaction reported in Scheme 60 [132]. For the synthesis of the silyl-functionalized PIB the allyl terminated polymer was hydrosilylated with dimethylchlorosilane. Hydroxy-terminated PEG was subjected to a Williamson reaction with allyl bromide to provide the allyl-functionalized PEG. Using PEG terminated with allyl groups at both ends PEG-*b*-PIB-*b*-PEG triblock copolymers were prepared. Products having rather low molecular weight were obtained, through this procedure, however, SEC data concerning the molecular weight distribution of the samples were not provided in this study.

An interesting development regarding the synthesis of block copolymers involves the use of bifunctional or dual initiators, which are compounds capable of performing two mechanistically distinct polymerizations. The advantage of this procedure is that there is no need for intermediate transformation or activation steps. If the different initiation sites are equally active for the

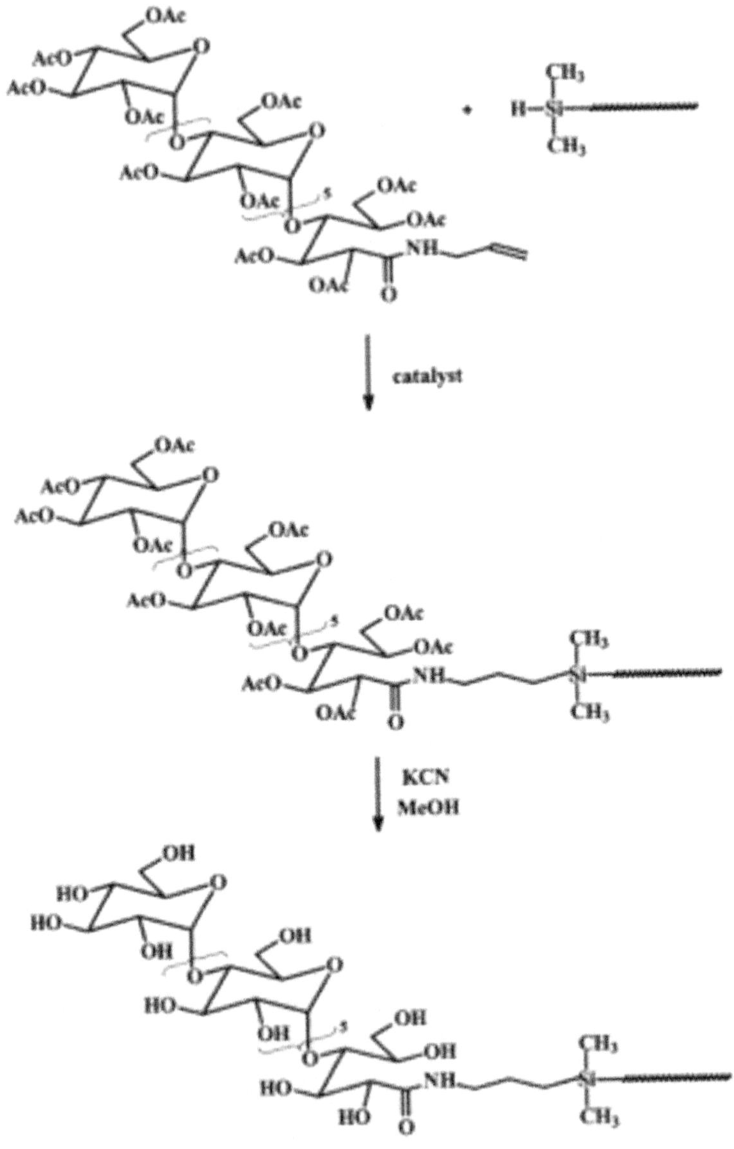

Scheme 58

polymerization of the various monomers, the synthesis of a wide range of block copolymers is both simple and efficient.

Sodium 4-oxy-2,2,6,6-tetramethyl-1-piperidinyloxy, TEMPONa, was used as a bifunctional initiator for the synthesis of PEO-*b*-PS block copolymers [133]. Initially the ROP of EO was performed in THF at 60 °C to provide narrow molecular weight distribution chains with terminal TEMPO moieties. Using these functionalized PEO chains the polymerization of styrene was

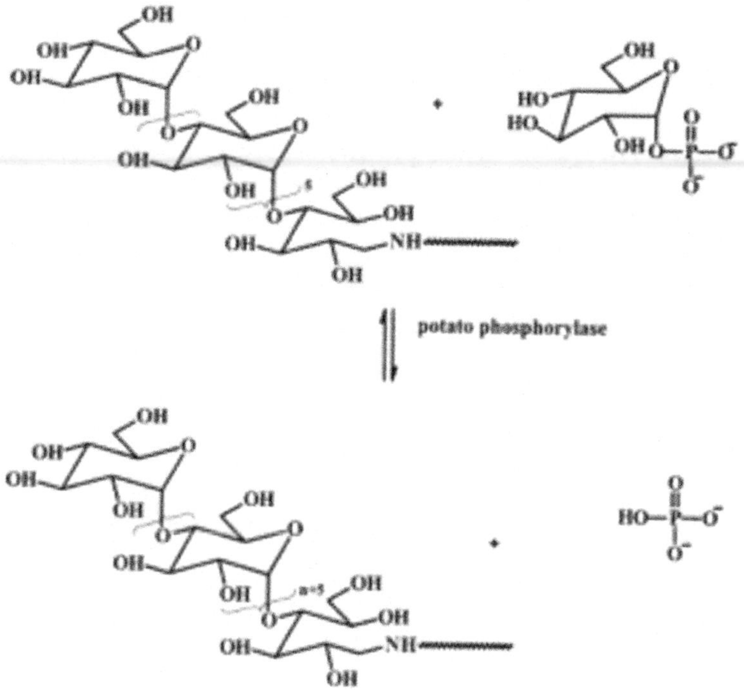

Scheme 59

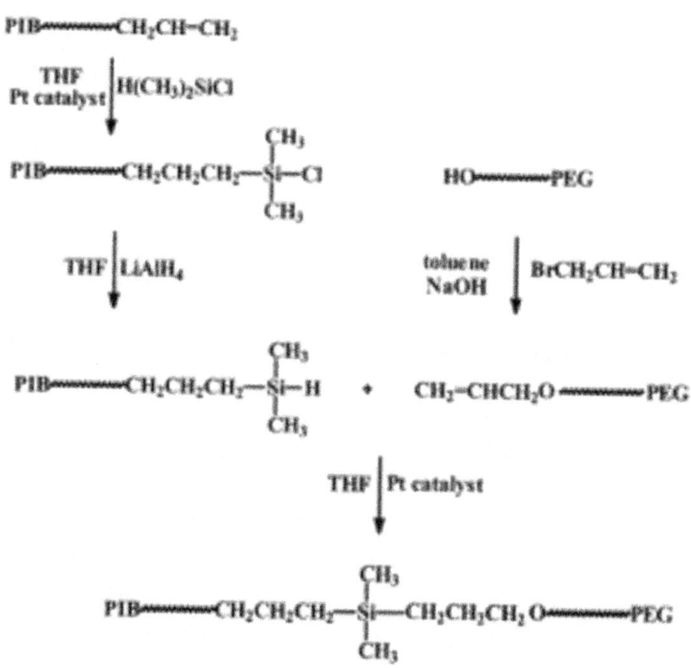

Scheme 60

carried out in the presence of AIBN at 120 °C (Scheme 61). Moderate polydispersities equal to 1.4 were obtained for the final products.

The bifunctional initiator 4-hydroxy-butyl-2-bromoisobutyrate, HBBIB, promoted the ATRP of styrene as well as the cationic ring opening polymerization of THF [134]. In the presence of $Cu/CuBr_2/PMDETA$ styrene was polymerized through the bromoisobutyrate function of HBBIB, to give PS chains end-functionalized with hydroxyl groups, PS – OH. The *in situ*

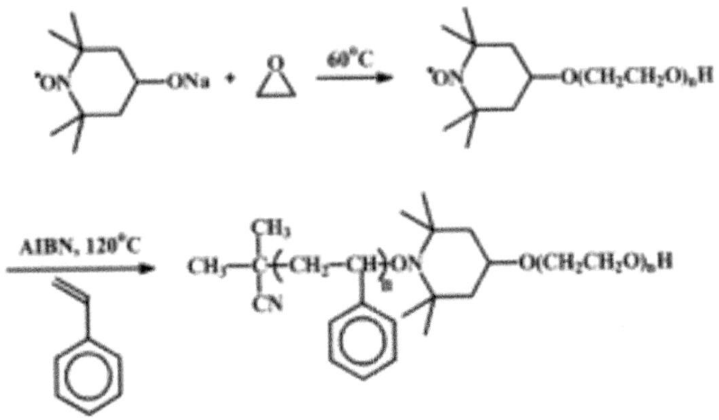

Scheme 61

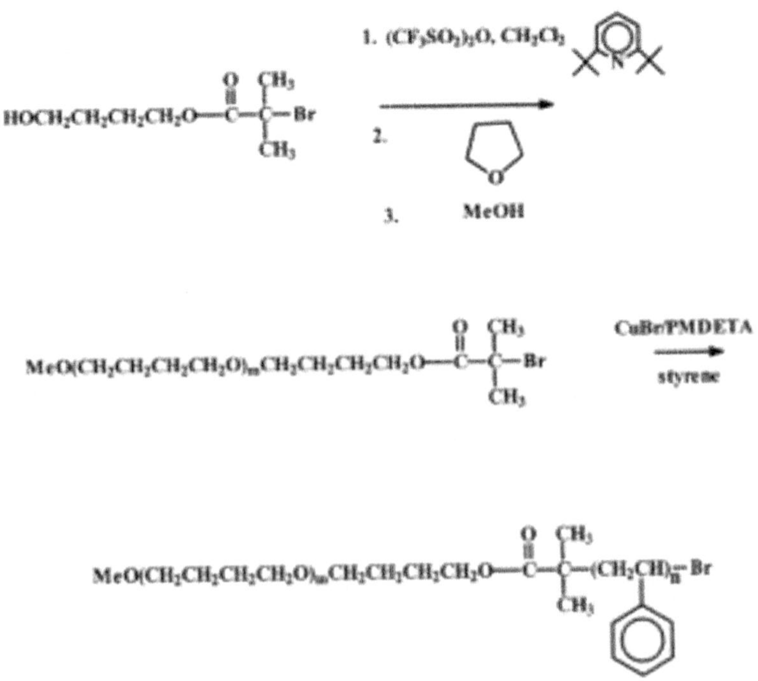

Scheme 62

reaction of the hydroxyl groups of HBBIB with trifluoromethane sulfonic anhydride provided a triflate ester group able to initiate the cationic ring opening polymerization of THF, leading to PTHF with terminal bromoisobutyrate groups, PTHF-Br. By starting the polymerization of THF with PS-OH, a bimodal distribution, consisting of the desired block copolymer and the PS homopolymer, was observed. On the contrary, with PTHF-Br as the macroinitiator for the polymerization of styrene, well-defined products were obtained (Scheme 62).

2-Phenyl-2-[(2,2,6,6-tetramethylpiperidino)oxy]ethyl 2-bromo-2-methyl propanoate and 2-phenyl-2-[(2,2,6,6-tetramethylpiperidino)oxy]ethyl 2-bromo-2-propanoate were utilized as dual initiators to promote the ATRP and the NMP of different monomers for the synthesis of block copolymers [135]. MMA or *t*BuA were polymerized in the presence of CuCl/PMDETA at 90 °C to afford the corresponding polymers with TEMPO terminal groups. Subsequent bulk polymerization of styrene using the PMMA or P*t*BuA macroinitiators at 125 °C yielded the final PMMA-*b*-PS (Scheme 63) or P*t*BuA-*b*-PS block copolymers. Moderate conversions were obtained for the polymerization of both the first and the second block, indicating poor control over molecular weight. Furthermore, moderate polydispersities, (between 1.3 and 1.5) were obtained.

The same heterobifunctional initiator, 2-phenyl-2-[(2,2,6,6-tetramethypiperidino)oxy]ethyl 2-bromo-2-methyl propanoate, was employed for the synthesis of PMMA-*b*-P*t*BuA-*b*-PS triblock terpolymers via the combination of ATRP and NMP [136]. Styrene was initially polymerized through the alkoxyamine function in bulk at 125 °C, leading to PS chains with bromine end groups. Subsequent addition of *t*BuA in the presence of CuBr/PMDETA provided the PS-*b*-P*t*BuA diblock. Further addition of CuCl, to achieve halogen exchange and MMA yielded the desired triblock copolymer with

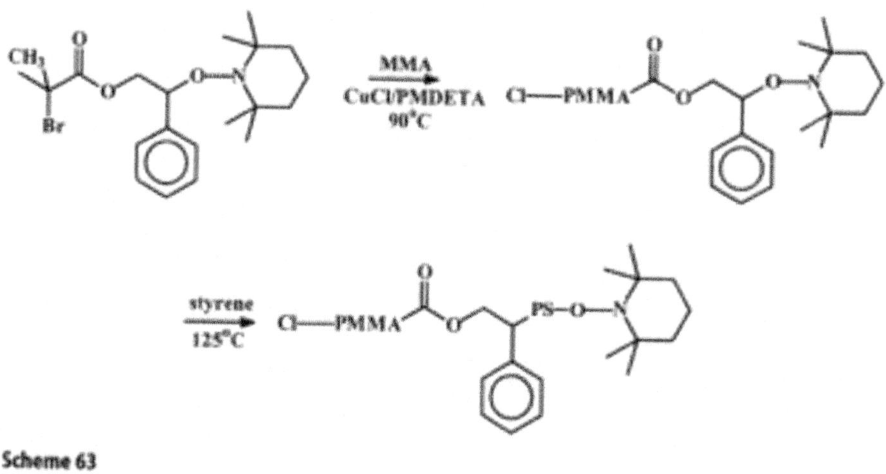

Scheme 63

a narrow molecular weight distribution but in rather low conversions. Attempts to first prepare the P*t*BuA-*b*-PMMA followed by the NMP of styrene resulted in tailing effects at the SEC trace, meaning that either termination reactions took place or that the TEMPO unit was not 100% efficient in promoting the polymerization of styrene in the presence of the P*t*BuA-*b*-PMMA block.

Bifunctional bipyridine initiators able to promote the ATRP and ROP have been used for the synthesis of a wide variety of block copolymers, such as PS-*b*-PMMA, PS-*b*-PCL, PMMA-*b*-PEG, PLA-*b*-PEG, PCL-*b*-PEG, and PCL-*b*-PMMA [137]. Starting from 4,4′-bis(chloromethyl)-2,2′-bipyridine a series of bipyridine derivatives carrying hydroxyl-, chlorine or bromine groups were synthesized, as illustrated in Scheme 64. The halogens were able to initiate the copper-catalyzed ATRP, whereas the hydroxyl groups promoted the aluminum-catalyzed ROP. Well-defined block copolymers of narrow molecular weight distributions were obtained through this methodology.

A bifunctional initiator combining the enzymatic ring opening polymerization of CL and the ATRP of styrene was utilized for the synthesis of PCL-*b*-PS block copolymers [138]. The initiator was prepared from the benzyl ester of bis(hydroxy)-propionic acid. One of the hydroxy groups was esterified with 2-bromo-2-methylpropionyl bromide. The primary alcohol group was used to initiate the enzymatic polymerization of CL using Novozym 435, a lipase immobilized on an acrylic acid, followed by the ATRP of styrene using the CuBr/PMDETA catalytic system. This order was chosen, since the enzymatic macroinitiator is not very efficient, due to the increased steric demands associated with this type of polymerization. In addition to polymerization, lipases catalyze the transesterification reactions leading to polymer degradation. To

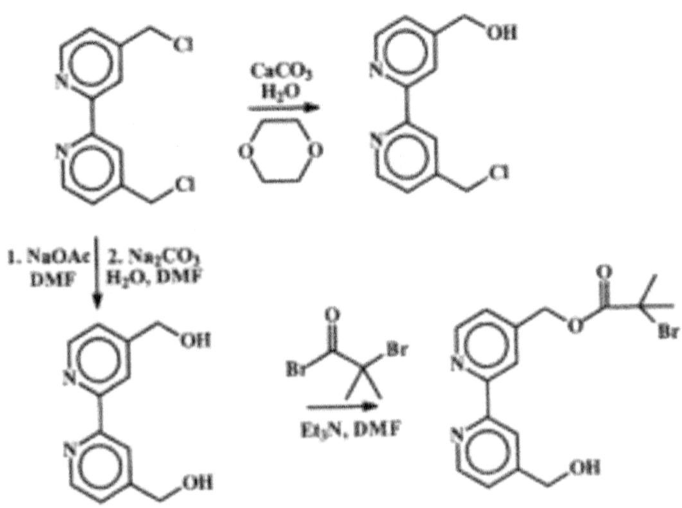

Scheme 64

avoid this side reaction, prolonged polymerization times were avoided. SEC analysis revealed that this procedure is efficient in promoting the synthesis of block copolymers. However, the molecular weight distribution, especially of the first enzymatically synthesized block, was rather broad.

3 Synthesis of Linear Multiblock Copolymers

Multiblock copolymers are linear copolymeric structures consisting of repeating units of a certain block copolymer of the type (A-*b*-B)$_n$. Here, A and B are macromolecular chains, usually of low molecular weight and n is the degree of polymerization of the copolymeric structure. The synthetic strategy used for the preparation of multiblock copolymers involves the synthesis of the individual A and B chains with functional groups such as hydroxyls and carboxyls at both ends. The functionalized chains are subsequently subjected to step growth polymerization for the preparation of the multiblock copolymer. For the synthesis of the difunctional A and B chains, living polymerization methods are usually employed, leading to controlled molecular weights, low polydispersities, and very high degrees of functionalization. However, the coupling of the A-*b*-B copolymeric chains suffer the drawbacks of step growth polymerization, where control over the degree of polymerization is difficult to achieve and the molecular weight distributions are high. Nevertheless, these materials possess interesting properties both in solution and in bulk.

Multiblock copolymers (P2VP-*b*-PEO)$_n$ were prepared by the condensation reaction of dihydroxy-terminated P2VP and PEO in dichloromethane in the presence of potassium hydroxide [139]. The difunctional P2VP was synthesized by anionic polymerization using the difunctional initiator lithium α-methylnaphthalene, followed by the end-capping reaction with EO. The products were contaminated with homopolymers, and thus extractions were necessary to obtain the pure copolymers. The molecular characterization was very poor since there were no details regarding the molecular weights and the molecular weight distributions.

(PLLA-*b*-PCL)$_n$ multiblock copolymers were prepared from the coupling reaction between the bischloroformates of carboxylated PLLA with diol-terminated PCL in the presence of pyridine [140]. LLA was polymerized with $SnOCt_2$ and 1,6-hexanediol followed by the reaction with succinic anhydride to provide the dicarboxylated PLLA. The carboxyl end groups were subsequently transformed to acid chloride groups by the reaction with thionyl chloride (Scheme 65). As expected, the molecular weight distributions were broad for all samples ($1.84 < M_w/M_n < 3.17$).

Multiblock copolymers of poly(L-lysine) and PEG were prepared following the reaction sequence illustrated in Scheme 66. *N*-carboxy-(N^ε-benzyloxy

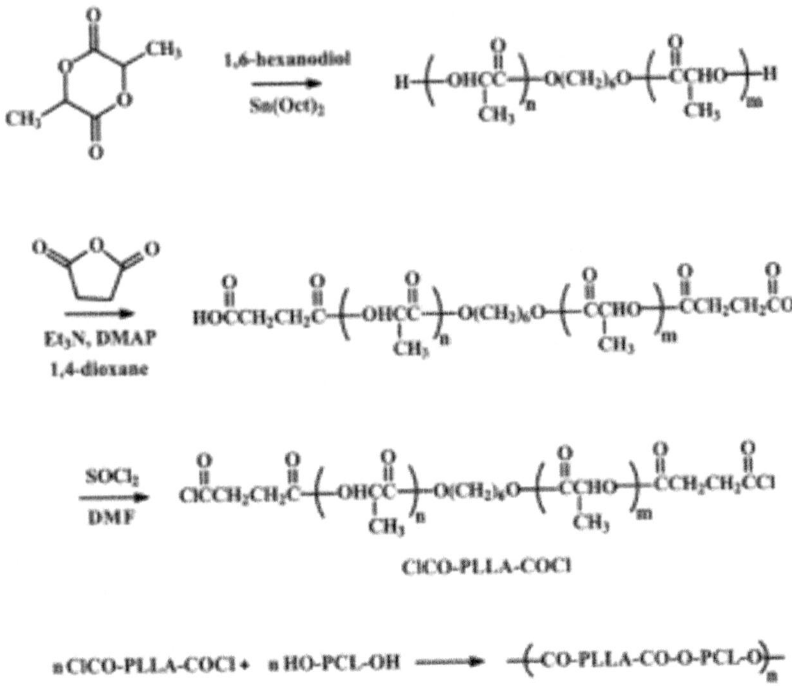

Scheme 65

carbonyl)-L-lysine anhydride was polymerized using the difunctional initiator 1,2-ethylenediamine [141]. This product was then copolymerized with PEG end-functionalized with succinimidyl succinate groups to provide the carbobenzoxy protected multiblock copolymer. Deprotection was carried out with formic acid in DMF using palladium catalyst. The number of PLL-*b*-PEG repeating units was higher than 5 in all cases. Molecular weight distributions of the copolymers were, however, very broad (M_w/M_n higher than 3.5).

4
Synthesis of Non-Linear Block Copolymers

4.1
Synthesis of Star-Block Copolymers

Star-block copolymers are star polymers in which each arm is a block (diblock or triblock) copolymer. There are several methods used for the synthesis of star-block copolymers [142], and the most commonly used strategies are given in Scheme 67.

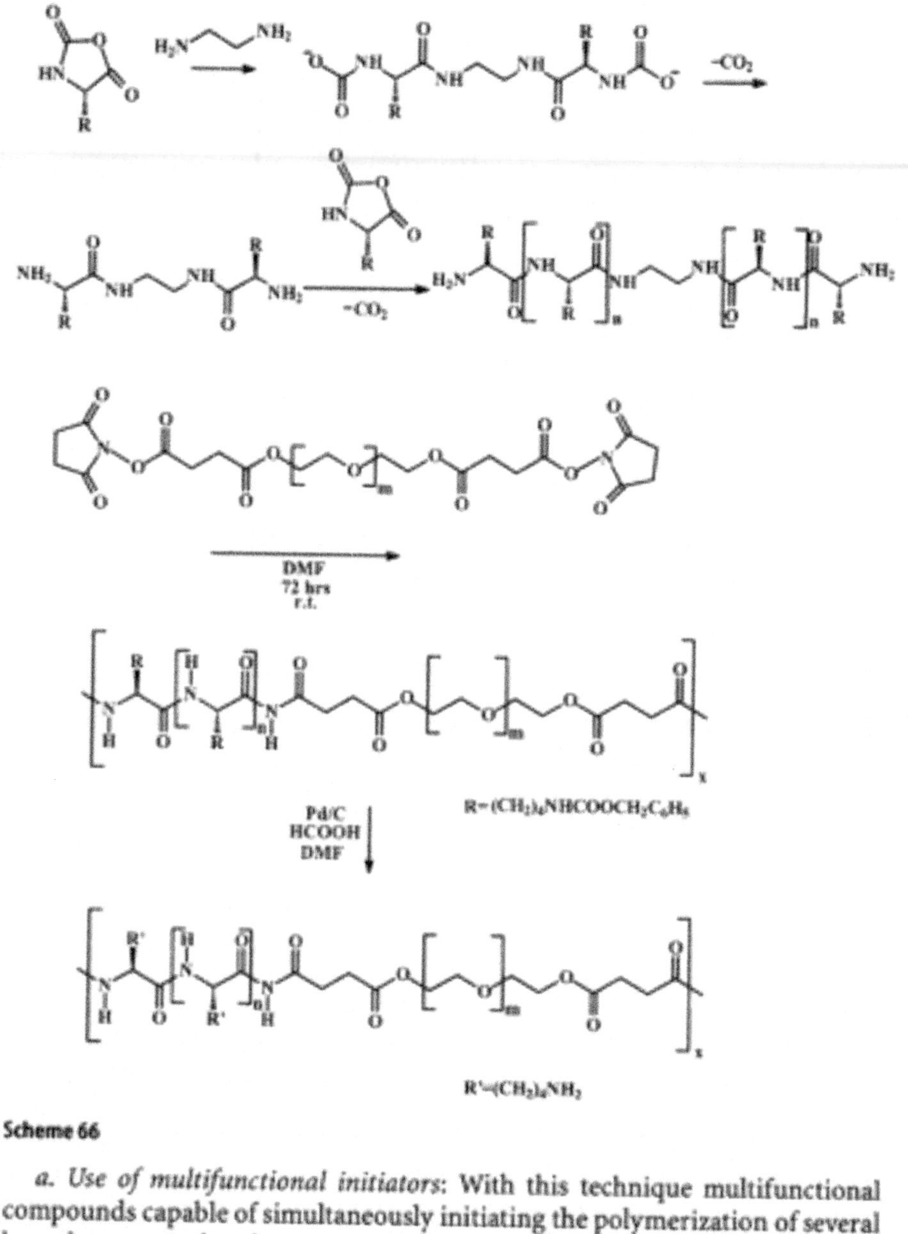

Scheme 66

a. Use of multifunctional initiators: With this technique multifunctional compounds capable of simultaneously initiating the polymerization of several branches are used to form a star polymer, A_n, where n is the functionality of the star. These living ends can then initiate the polymerization of the second monomer to give the star-block copolymer, $(A\text{-}b\text{-}B)_n$ or they can react with the end-functionalized pre-synthesized B chains to afford the same product. Several requirements are necessary for a multifunctional initiator to produce star polymers with uniform arms, low molecular weight distribution and controllable molecular weights. All the initiation sites must be equally reactive

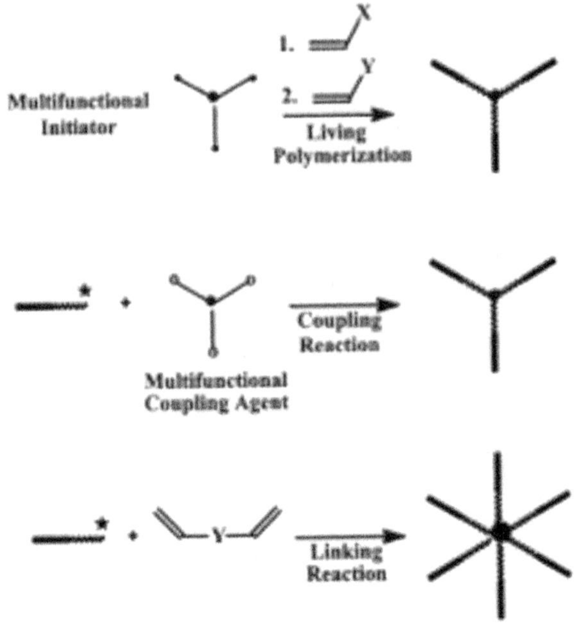

Scheme 67

and have the same rate of initiation. Furthermore, the initiation rate must be higher than the propagation rate. Steric hindrance repulsions, caused by the high segment density, result in excluded volume effects.

b. Use of multifunctional linking agents: This method involves the synthesis of living block copolymeric chains, using methods previously reported and their subsequent reaction with a multifunctional linking agent. The absolute control in all the synthetic steps renders this technique the most efficient for the synthesis of well-defined star polymers. The functionality of the linking agent determines the number of branches in the star polymer, provided that the linking reaction is quantitative. The living arms can be isolated and characterized independently along with the final star product. Consequently, the star's functionality can be measured directly and accurately. Disadvantages of the method include the long time occasionally required for the linking reaction, and the need to perform fractionation in order to obtain the pure star polymer, since in almost all cases a small excess of the living arm is used in order to assure complete linking.

c. Use of difunctional monomers: With this strategy a living block copolymer precursor is used as the initiator for the polymerization of a small amount of a suitable difunctional monomer. Microgel nodules of tightly crosslinked polymer are formed upon polymerization. These nodules serve as the branch point from which the arms emanate. The functionality of the stars prepared by this method can be determined by molecular weight measurements on the arms and the star product but prediction and control of arm number is very difficult. The number of branches incorporated in the star structure is influenced by many parameters, the most important being the molar ratio of the difunctional monomer over the living polymer. The functionality of the star increases by increasing this ratio. Other parameters which influence the number of branches include the chemical nature of the copolymers, the concentration and the molecular weight of the living copolymer chain, the temperature and the time of the reaction, the rate of stirring etc. Another disadvantage of this procedure is that the final products are characterized by a distribution in the number of arms incorporated into the star structure. Consequently, the number of arms determined experimentally by molecular weight measurements is an average value. It is obvious that although this method is technologically very important and can be applied on an industrial scale it is less suitable for the preparation of well-defined stars.

Other less common methods for the synthesis of star-block copolymers have also been reported. Recent characteristic examples will be given in the following paragraphs.

Amphiphilic star-block copolymers with four arms, {polycaprolactone-*b*-poly[4-(2-hydroxyethyl)caprolactone]}$_4$ were prepared. Pentaerythritol was used as a tetrafunctional initiator in the presence of tin 2-ethylhexanoate to produce the polycaprolactone four arm star [143]. SEC with light scattering detector and NMR measurements showed that well-defined near monodisperse low molecular weight stars were prepared. These 4-arm stars were subsequently used as macroinitiators for the polymerization of 4-(2-benzyloxyethyl)caprolactone in order to obtain the corresponding star-block copolymers. The polymerization was conducted at 110 °C in the presence of the catalyst tin 2-ethylhexanoate (Scheme 68). The copolymerization was successful, judging from the SEC traces. However, the molecular weight distributions broadened upon copolymerization. The benzyl masking groups were removed by catalytic debenzylation in the presence of H_2 and Pd/C catalyst.

[Poly(isobutylene)-*b*-polynorbornadiene]$_3$, (PIB-*b*-PNBD)$_3$ (Scheme 69) with either PIB or PNBD as inner blocks were prepared by cationic polymerization techniques [144]. Tricumylchloride and $TiCl_4$ were used to provide a trifunctional initiation system to promote the living polymerization of IB and lead to the synthesis of the corresponding 3-arm star. The polymerization was conducted in $CH_3Cl/CHCl_3$ 30/70 (v/v) at − 35 °C to achieve the best conditions for obtaining well-defined stars. Subsequent addition of NBD yielded the desired star-block copolymers. Due to chain transfer reactions the product was contaminated with PNBD homopolymer. In addition, star-star coupling was also observed by SEC analysis. This was attributed to the fact that a growing arm may cationate an arm of another star with a $-CH_2-C(CH_3)=CH_2$ end group, previously formed by chain transfer. Using the reverse mode of monomer addition, NBD was first polymerized followed by the addition of IB. In this case the molecular weight distribution was

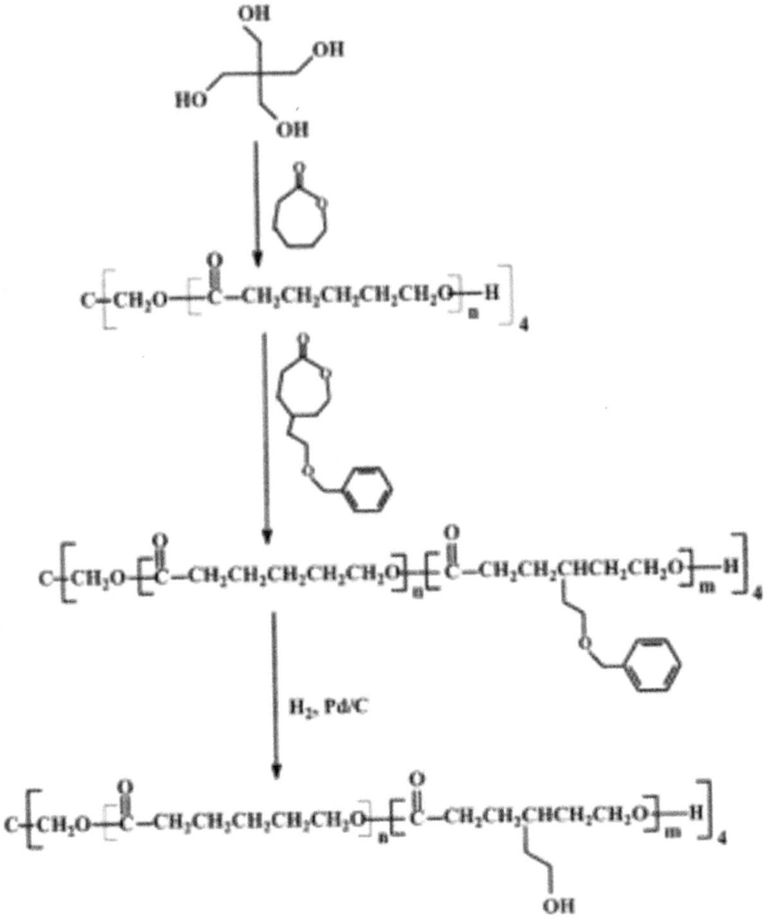

Scheme 68

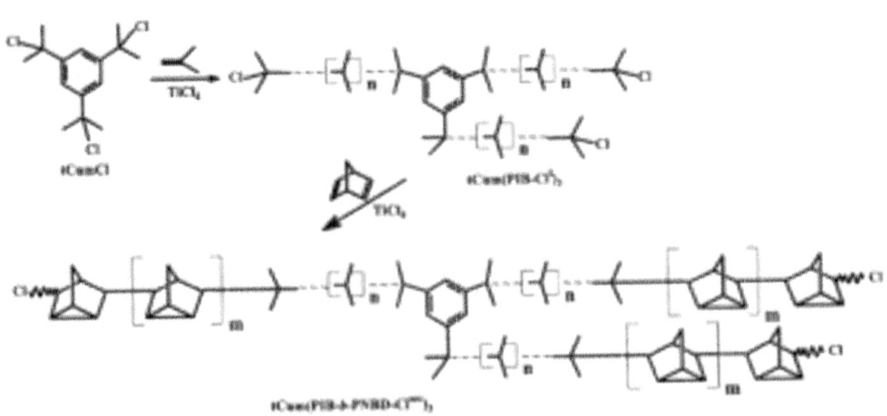

Scheme 69

broad ($M_w/M_n \sim 2$) and similar by-products were obtained as in the previous case.

Hexaepoxy squalene, HES (Scheme 70) was used as a multifunctional initiator in the presence of $TiCl_4$ as a coinitiator, di-*t*-butylpyridine as a proton trap, and *N,N*-dimethylacetamide as an electron pair donor in methylcyclohexane/methyl chloride solvent mixtures at −80 °C for the synthesis of (PIB-*b*-PS)$_n$ star-block copolymers [145]. IB was polymerized first followed by the addition of styrene. The efficiency and the functionality of the initiator were greatly influenced by both the HES/IB ratio and the concentration of $TiCl_4$, thus indicating that all epoxy initiation sites were not equivalent for polymerization. Depending on the reaction conditions stars with 3 to 10 arms were synthesized. The molecular weight distribution of the initial PIB stars was fairly narrow ($M_w/M_n < 1.2$), but it was sufficiently increased after the polymerization of styrene ($1.32 < M_w/M_n < 1.88$).

The oxocarbenium perchlorate $C(CH_2OCH_2CH_2CO^+ClO_4^-)_4$ was employed as a tetrafunctional initiator for the synthesis of PTHF 4-arm stars [146]. The living ends were subsequently reacted either with sodium bromoacetate or bromoisobutyryl chloride. The end-capping reaction was not efficient in the first case (lower than 45%). Therefore, the second procedure was the method of choice for the synthesis of the bromoisobutyryl star-shaped macroinitiators. In the presence of CuCl/bpy the ATRP of styrene was initiated in bulk, leading to the formation of (PTHF-*b*-PS)$_4$ star-block copolymers. Further addition of MMA provided the (PTHF-*b*-PS-*b*-PMMA)$_4$ star-block terpolymers. Relatively narrow molecular weight distributions were obtained with this synthetic procedure.

1,3,5-Benzenetricarbonyl trichloride and 1,2,4,5-tetrakis(bromomethyl) benzene were employed as multifunctional initiators for the synthesis of 3- and 4-arm PTHF stars, respectively [147]. The living ends were reacted with sodium 2-bromoisobutyrate followed by reduction with SmI_2. The samarium enolates, thus formed were efficient initiators for the polymerization of MMA to give the (PTHF-*b*-PMMA)$_n$, $n = 3$ or 4 star-block copolymers, according to Scheme 71.

The relatively slight broadening of the molecular weight distribution during the polymerization of the MMA blocks showed that the macroinitiator (PTHF)$_4$ is highly efficient in promoting the block copolymerization.

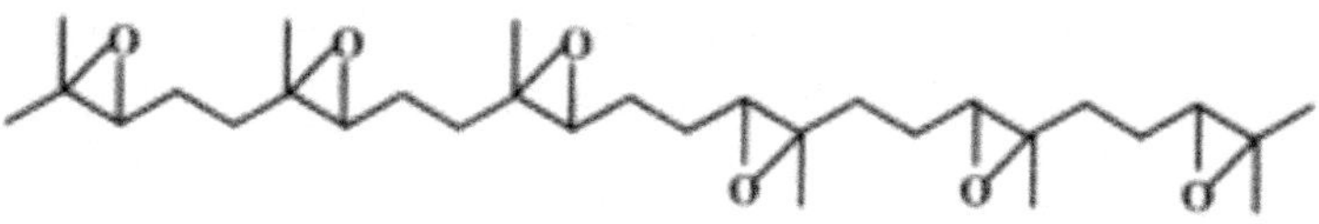

HES

Scheme 70

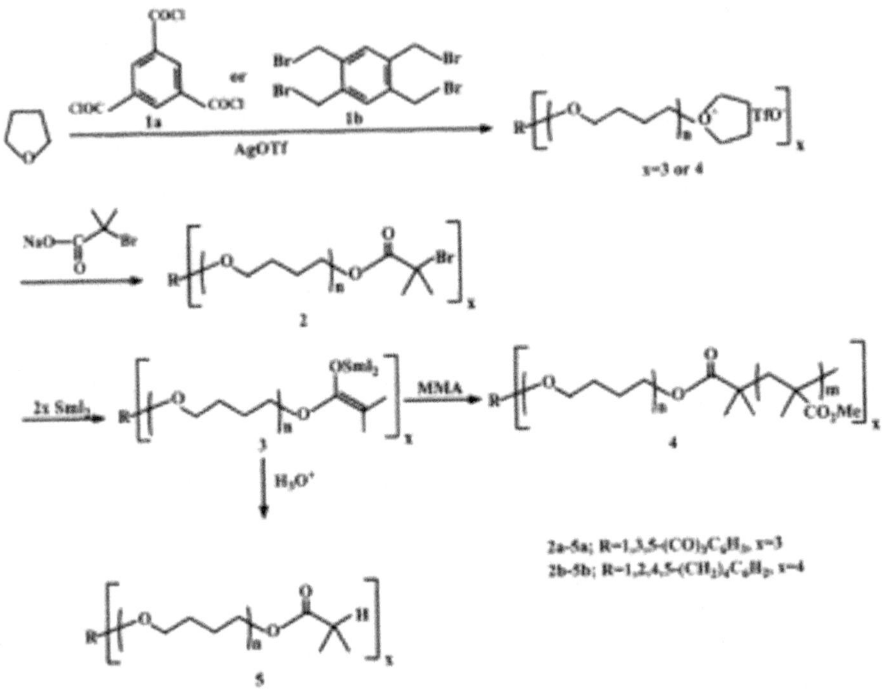

Scheme 71

A combination of anionic and ATRP was employed for the synthesis of (PEO-*b*-PS)$_n$, $n = 3$, 4 star-block copolymers [148]. 2-Hydroxymethyl-1,3-propanediol was used as the initiator for the synthesis of the 3-arm PEO star. The hydroxyl functions were activated by diphenylmethyl potassium, DPMK in DMSO as the solvent. Only 20% of the stoichiometric quantity of DPMK was used to prevent a very fast polymerization of EO. Employing pentaerythritol as the multifunctional initiator a 4-arm PEO star was obtained. Well-defined products were provided in both cases. The hydroxyl end groups of the star polymers were activated with DPMK and reacted with an excess of 2-bromopropionyl bromide at room temperature. Using these 2-bromopropionate-ended PEO stars in the presence of CuBr/bpy the ATRP of styrene was conducted in bulk at 100 °C, leading to the synthesis of the star-block copolymers with relatively narrow molecular weight distributions (Scheme 72).

(PS-*b*-PEO)$_n$, $n = 3$, 4 star-block copolymers were synthesized by ATRP and anionic polymerization techniques [149]. Three- or four-arm PS stars were prepared using tri- or tetrafunctional benzylbromide initiators in the presence of CuBr/bipy. The polymerization was conducted in bulk at 110 °C. The end bromine groups were reacted with ethanolamine in order to generate the PS stars with hydroxyl end groups. These functions were then activated by DPMK to promote the polymerization of ethylene oxide and afford the desired well-defined products (Scheme 73).

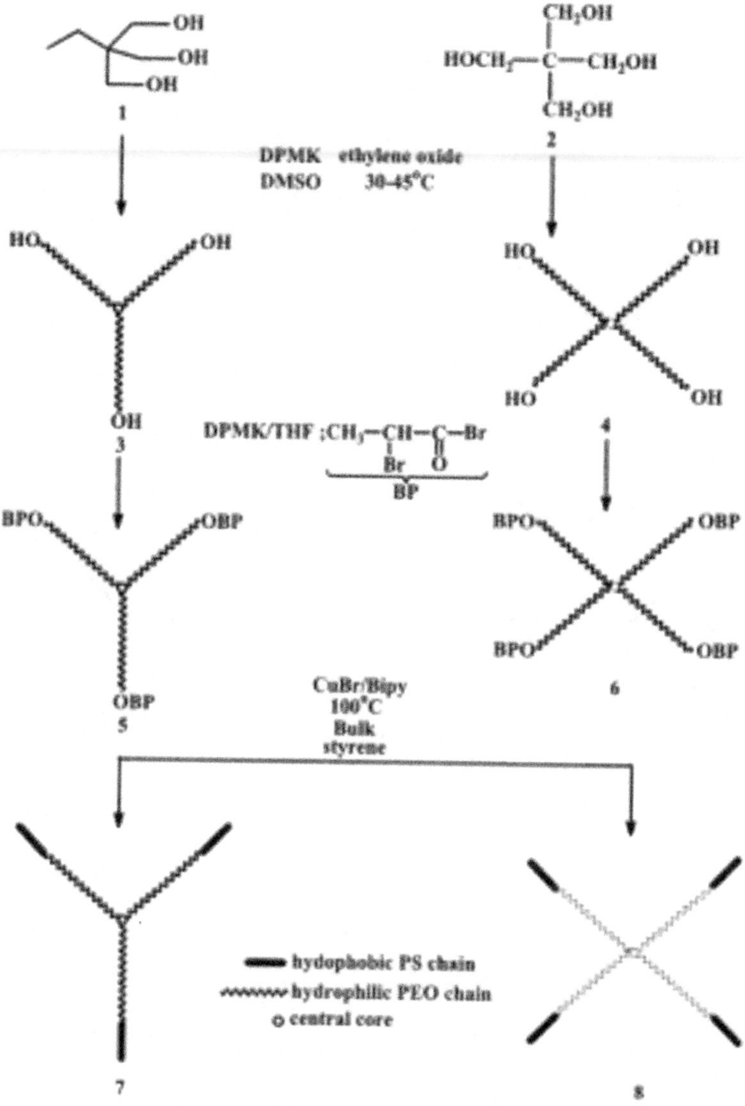

Scheme 72

Double hydrophilic star-block (PEO-*b*-PAA)$_3$ copolymers were prepared by a combination of anionic and ATRP of EO and *t*BuA [150]. Three-arm PEO stars, with terminal – OH groups were prepared by anionic polymerization, using 1,1,1-tris(hydroxymethyl)ethane, activated with DPMK as a trifunctional initiator. The hydroxyl functions were subsequently transformed to three bromo-ester groups, which were utilized to initiate the polymerization of *t*-butyl acrylate by ATRP in the presence of CuBr/PMDETA. Subsequent hydrolysis of the *t*-butyl groups yielded the desired products (Scheme 74).

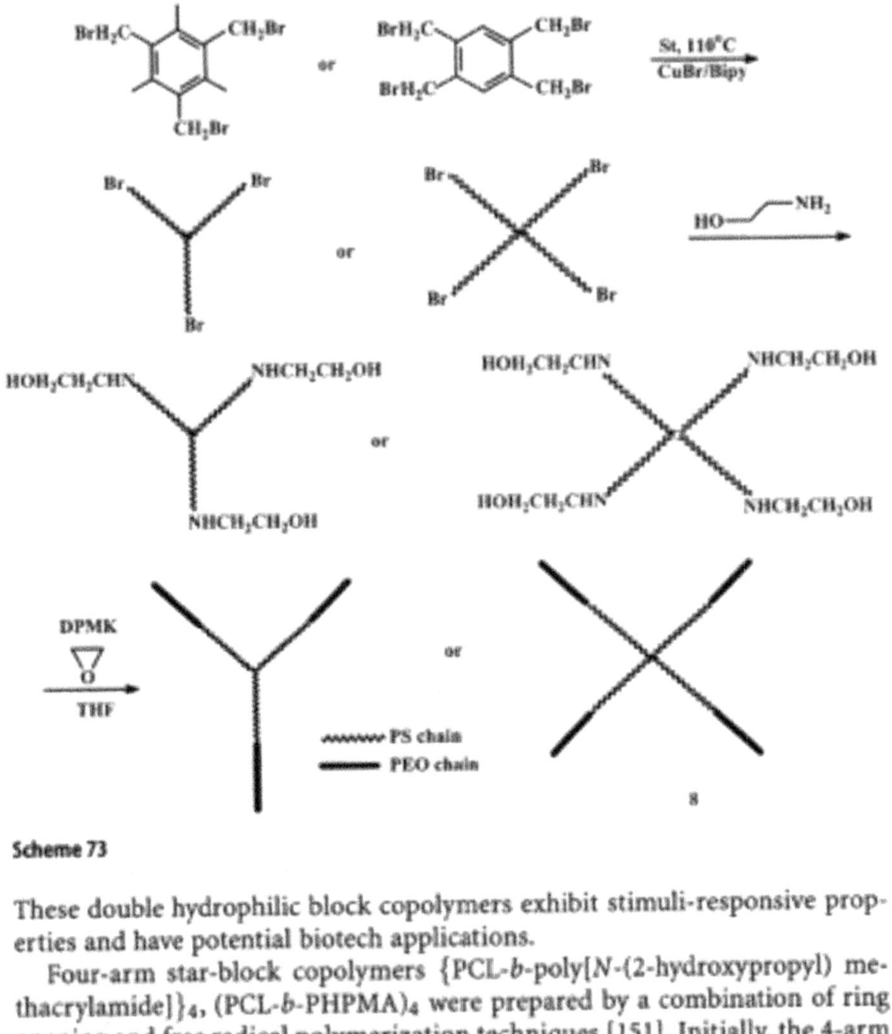

8

Scheme 73

These double hydrophilic block copolymers exhibit stimuli-responsive properties and have potential biotech applications.

Four-arm star-block copolymers {PCL-*b*-poly[*N*-(2-hydroxypropyl) methacrylamide]}$_4$, (PCL-*b*-PHPMA)$_4$ were prepared by a combination of ring opening and free radical polymerization techniques [151]. Initially, the 4-arm PCL star was prepared using pentaerythritol as the initiator and stannous 2-ethyl hexanoate as the catalyst at 150 °C. The end-hydroxyl groups were reacted with 3,3′-dithiobis(propionic acid), DTPA, using dicyclohexyl carbodiimide, DCC. Crosslinking was observed to some extent, due to the reactivity of the hydroxyl end groups. The star polymers with the thiol end groups were obtained after dithiothreitol-mediated reduction of the disulfide bonds. HPMA was then polymerized using AIBN as the initiator in the presence of the P(CL-SH)$_4$ star as the chain transfer agent (Scheme 75). The final products were characterized by broad molecular weight distributions ($M_w/M_n \sim 1.7$).

(PLLA-*b*-PEO)$_3$ star-block copolymers have been synthesized by a combination of ROP and post-polymerization reactions [152], as depicted in Scheme 76. Glycerol was employed for the synthesis of a 3-arm PLLA star

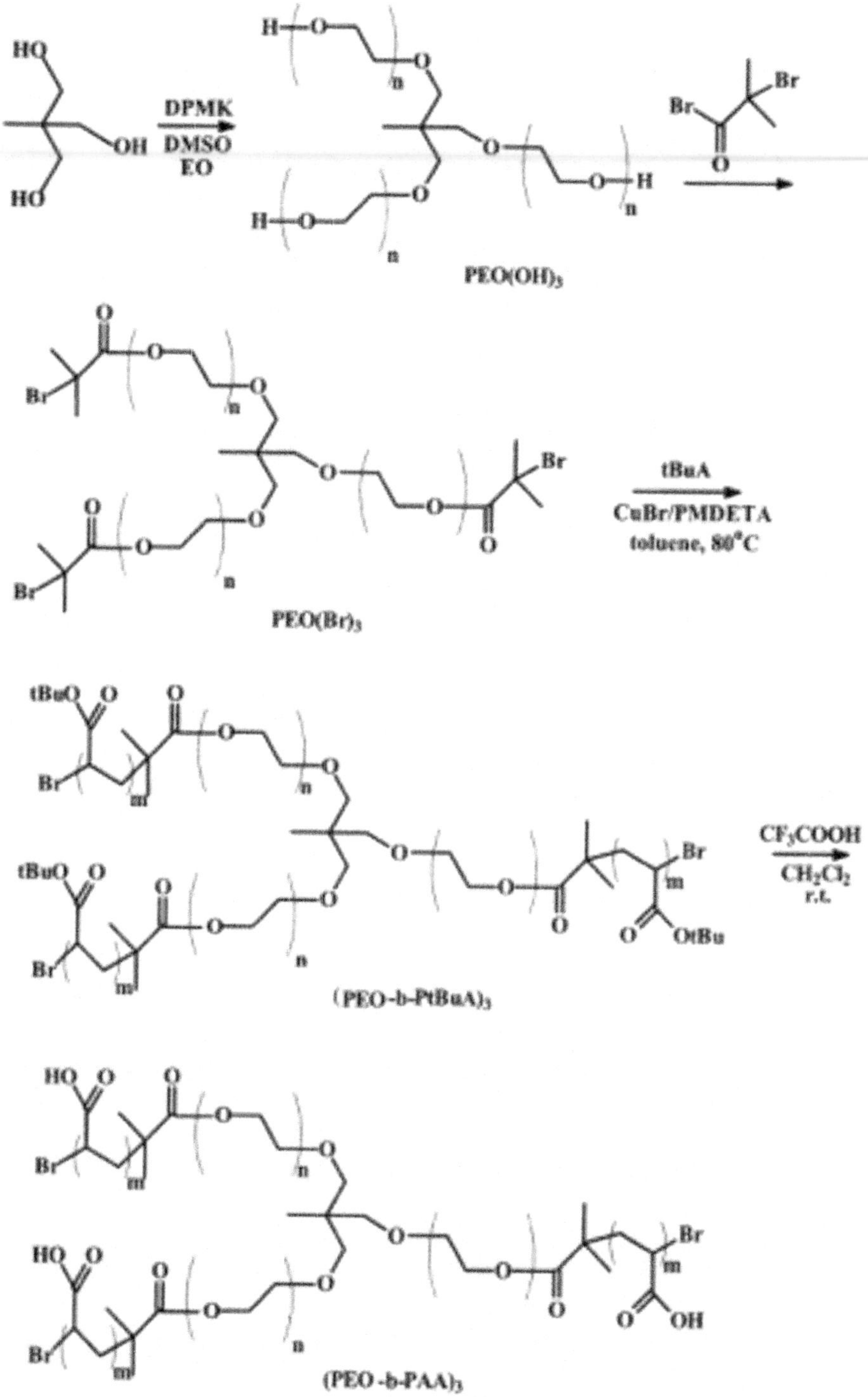
DPMK
DMSO
EO
PEO(OH)3
tBuA
CuBr/PMDETA
toluene, 80°C
PEO(Br)3
CF3COOH
CH2Cl2
r.t.
(PEO-b-PtBuA)3
(PEO-b-PAA)3

Scheme 74

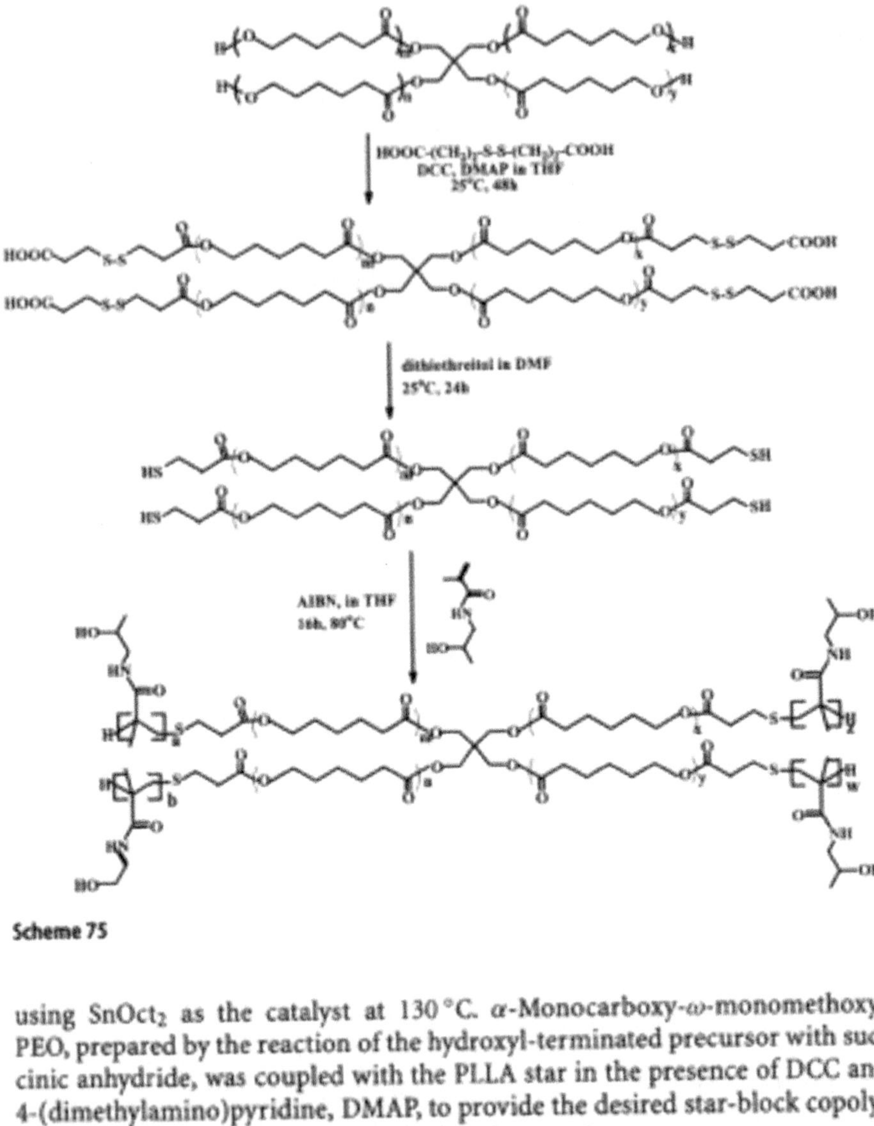

Scheme 75

using $SnOct_2$ as the catalyst at 130 °C. α-Monocarboxy-ω-monomethoxy-PEO, prepared by the reaction of the hydroxyl-terminated precursor with succinic anhydride, was coupled with the PLLA star in the presence of DCC and 4-(dimethylamino)pyridine, DMAP, to provide the desired star-block copolymers. NMR measurements showed that the reaction sequence was successful. Moderately broad molecular weight distributions ($1.18 < M_w/M_n < 1.32$) were obtained.

Anionic polymerization and suitable linking chemistry were employed for the synthesis of 3-arm PCHD-*b*-PS star-block copolymers with PCHD either as the inner or the outer block (Scheme 77) [153]. The block copolymers were prepared by sequential addition of monomers. It was shown that the crossover reaction of either PSLi or PCHDLi was efficient and led to well-defined block copolymers. However, in the case of the PCHD-*b*-PSLi copolymers, longer polymerization times were needed for long PCHD

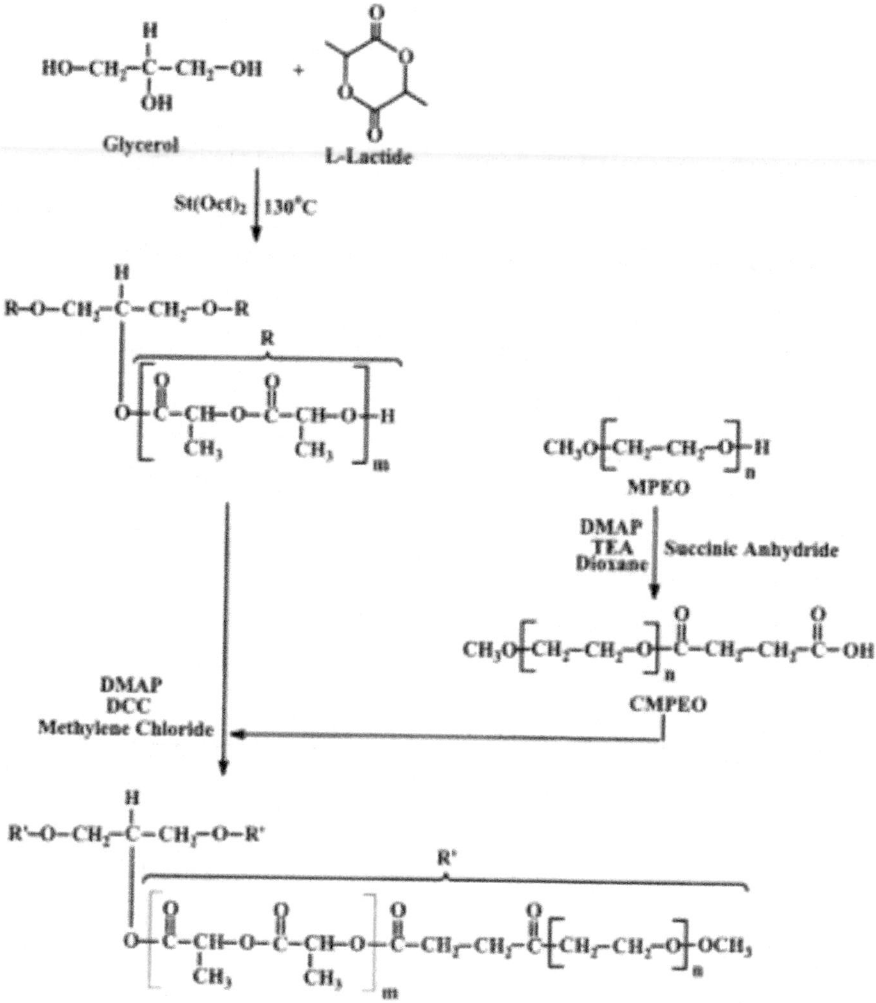

Scheme 76

blocks. During this interval termination and chain transfer reactions may take place to some extent, leading to contamination of the final product with PCHD homopolymer. CHD was polymerized in the presence of 1,4-diazabicyclo[2.2.2]octane, DABCO. The linking reaction was performed with CH_3SiCl_3 as the linking agent. When PCHD was the inner block the linking was efficient in spite of the long reaction times (up to 2 weeks). It was found that the linking reaction was facilitated by the presence of DABCO. In all cases, the living diblocks were end-capped with a few units of butadiene in order to reduce the steric hindrance of the living end and thus facilitate the linking process. Near monodisperse polymers were synthesized with this procedure.

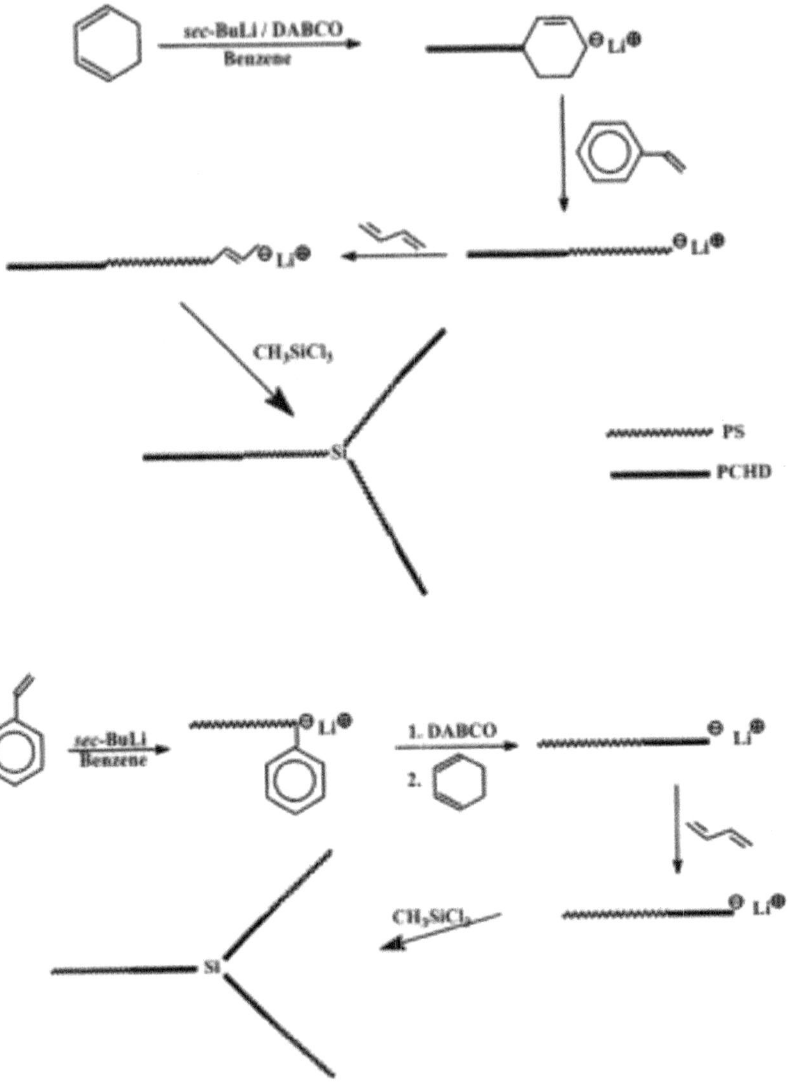

Scheme 77

The fullerene C_{60} was used as the linking agent for the synthesis of $(PCHD\text{-}b\text{-}PS)_6$ and $(PS\text{-}b\text{-}PCHD)_6$ star-block copolymers [154]. The polymers were then aromatized with 2,3-dichloro-5,6-dicyano-1,4-benzoquinone, DDQ, in 1,2-dichlorobenzene to yield the corresponding copolymers containing poly(1,4-phenylene) blocks. In order to achieve high 1,4-isomer contents and to avoid termination reactions, the polymerization of CHD was conducted in toluene at 10 °C without the presence of any additive to yield products with low molecular weights. Coupling of the PCHD-*b*-PSLi to C_{60}

was efficient. However, coupling of the reverse block copolymer PS-*b*-PCHDLi presented several difficulties, and subsequent aromatization led to degradation reactions. To avoid these problems the PS-*b*-PCHDLi chains were end-capped with a few units of styrene (Scheme 78).

PEO-*b*-P*t*BuA block copolymers were prepared by copper-mediated ATRP of *t*BuA using ω-brominated PEO macroinitiators. The polymerization was conducted at either 70 or 80 °C in toluene with the CuBr/PMDETA catalytic system. The living linear diblock precursors were then reacted with divinylbenzene in anisole and the same catalytic system to afford multiarm star-block copolymers (Scheme 79). The star polymer yield was found to range between 40 and 80% depending on the amount of DVB. The presence of the residual linear precursor can be explained by steric hindrance effects, loss of the halogen end group of the dormant species, existence of termination reactions etc. Maintaining the molar ratio of DVB and the block copolymer constant the molecular weight of the star was affected by the molecular weight of the linear precursor. The lower the molecular weight of the precursor the higher was the functionality of the star copolymer (ranging from 5 up to 82). Molecular weight distributions were rather broad ($1.26 < M_w/M_n < 1.93$).

A modular strategy for the preparation of functional multiarm star polymers with nitroxide-mediated "living" radical polymerization has been proposed [155]. The approach involves the use of a variety of alkoxyamine-functional initiators for the polymerization of several vinyl monomers. These linear chains, containing a dormant chain end, were coupled with a crosslinkable monomer, such as DVB, to yield a star polymer. The ability of this initiator to polymerize a great variety of vinyl monomers, along with the high diversity of the block sequence, led to the synthesis of a myriad of functionalized three-dimensional star polymers, such as (PS-*b*-P*t*BuA)$_n$, [PS-*b*-poly(*N*,*N*-dimethylacrylamide)]$_n$, (PS-*b*-PDMAA)$_n$, (PS-*b*-P2VP)$_n$, etc. A few examples are given in Scheme 80. No homopolymer contamination was detected after the synthesis of the stars. The number of the arms was calculated

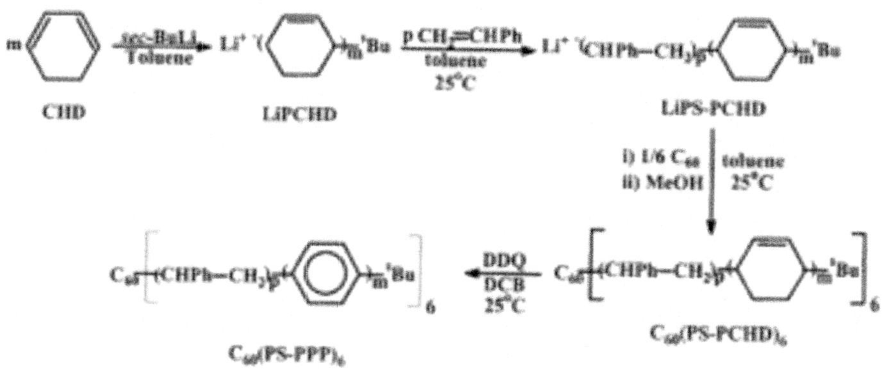

Scheme 78

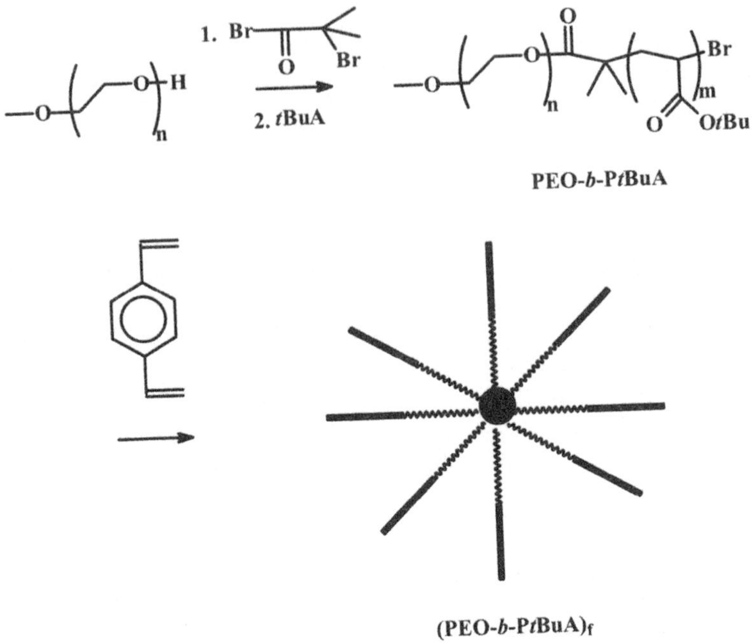

Scheme 79

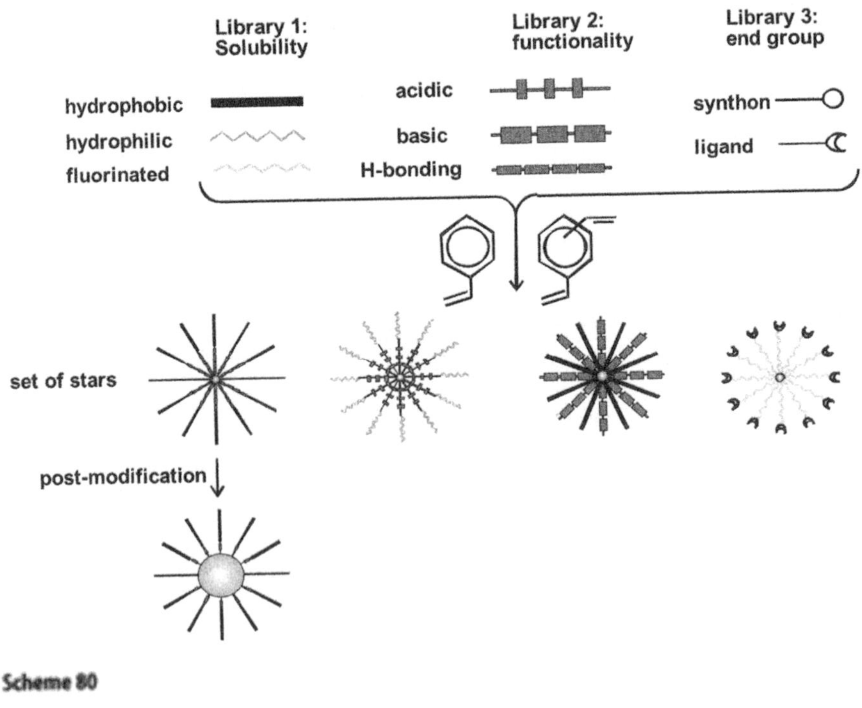

Scheme 80

to vary between 30 and 40. The versatility of the method increases when performing post-polymerization reactions on specific blocks of the stars. These unique structures are useful in a range of applications, such as supramolecular hosts, catalytic scaffolds, and as substrates for nanoparticle formation.

[PS-*b*-poly(4-*t*-butylstyrene)]*n*, $(PS\text{-}b\text{-}PtBuS)_n$ star-block copolymers were prepared by anionic polymerization and sequential addition of monomers with DVB as the linking agent for the formation of the star structure [156]. The functionality of the stars ranged between 10 and 20. Selective sulfonation of PS blocks was subsequently performed using the sulphur trioxide and triethyl phosphate complex in 1,2-dichloroethane, followed by neutralization with sodium methoxide. For this reason DVB was used for the linking reaction instead of chlorosilanes, where a better control could be achieved. DVB stars are more robust and the sulfonation reaction proceeds without cleavage of the arms from the star structure.

Ruthenium-catalyzed ATRP was employed in the synthesis of PMMA-*b*-P*n*BuMA block copolymers. Subsequent reaction with the divinyl compound 1 (Scheme 82) resulted in the synthesis of the star-block structures in almost quantitative yield [157]. The divinyl compound 2 was also employed for the linking of P*n*BuMA-*b*-PMMA through the PMMA blocks. Narrow molecular weight distribution products were obtained in all cases.

$(PS\text{-}b\text{-}PBd)_n$ star-block copolymers were synthesized by the macromonomer technique in combination with anionic polymerization and ROMP [158], following the procedure outlined in Scheme 83. The macromonomers were prepared with two different methods. In the first the living diblock copolymer was reacted with ethylene oxide to reduce the nucleophilicity of the living end followed by termination with 5-carbonyl chloride bicycle (2.2.1) hept-2-ene, while in the second method the functional initiator 5-lithiomethyl bicycle

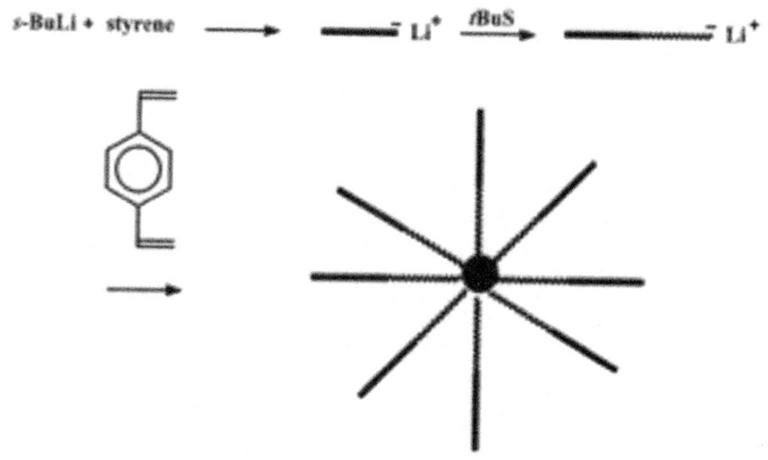

Scheme 81

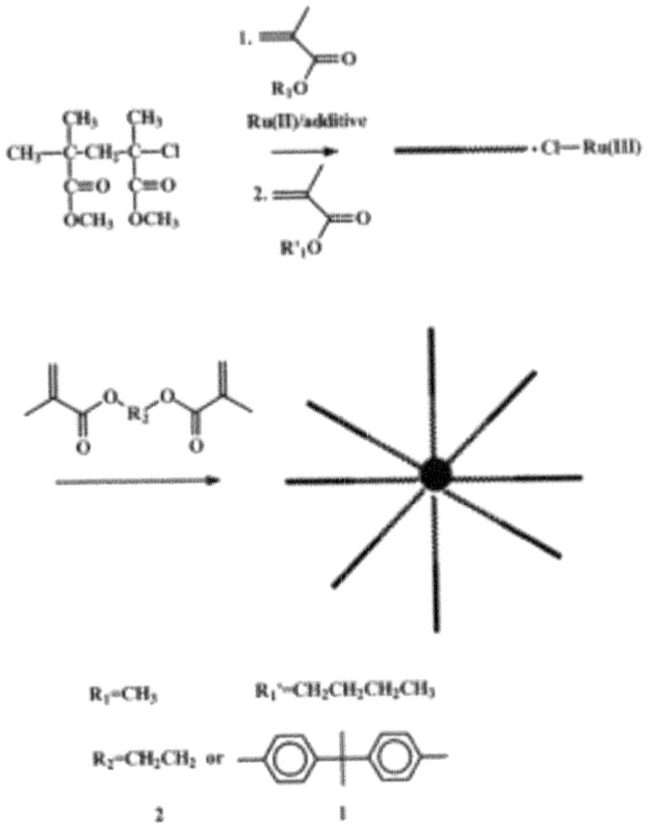

Scheme 82

(2.2.1) hept-2-ene was employed for the sequential polymerization of the two monomers. The first method places the norbornenyl group at the PB chain end, whereas the second one places the functional group at the PS chain end. Homopolymerization of these macromonomers took place with the catalyst Mo(NAr)(CHtBu) – (OtBu)$_2$. The polymerization conversion was very high (~ 90%) and no evidence of any degradation reaction was observed.

Bipyridine-centered triblock copolymers of the type BA-bpy-AB were prepared by a combination of ATRP and ROMP [159]. 4,4′-Bis(hydroxymethyl)-2,2′-bipyridine was employed for the polymerization of lactic acid, LA or CL in the presence of Sn(Oct)$_2$ in bulk at 130 and 110 °C, respectively. The hydroxyl end groups were converted to tertiary or secondary bromoesters by reaction with 2-bromoisobutyryl bromide or 2-bromopropionyl bromide. The reaction yields were very high (> 80%) but not quantitative. These products were used as macroinitiators for the ATRP of MMA or tBuA in the presence of CuBr/HMTETA. 4,4′-bis(Chloromethyl)-2,2′-bipyridine was employed to promote the ATRP of MMA or styrene followed by the addition

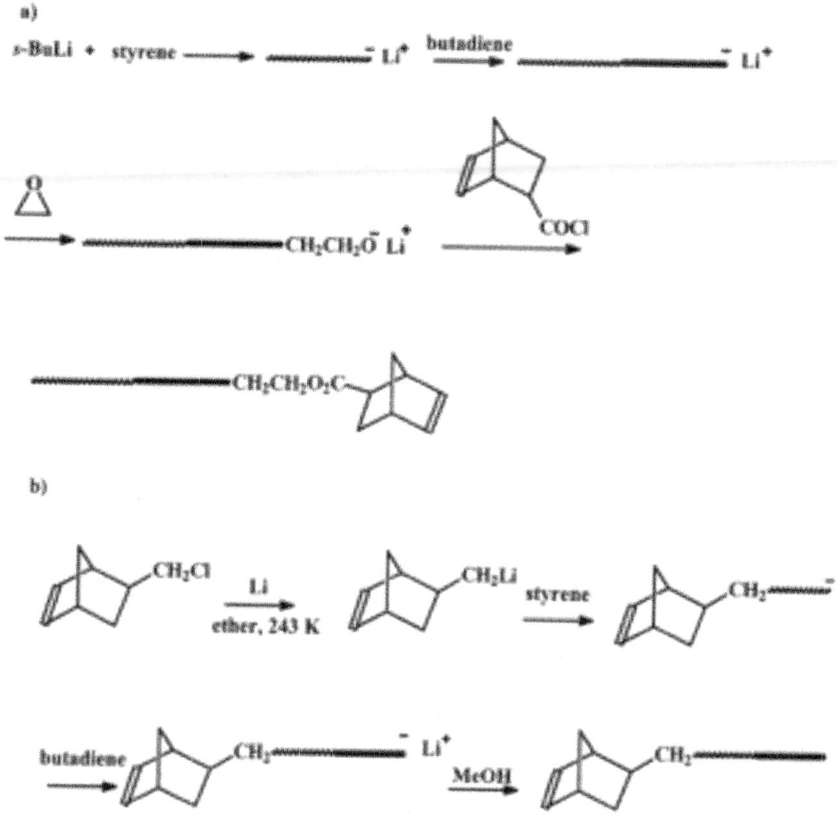

Scheme 83

of styrene or MMA or *t*BuA for the synthesis of PS-PMMA-bpy-PMMA-PS, PMMA-PS-bpy-PS-PMMA, and P*t*BuA-PS-bpy-PS-P*t*BuA triblock copolymers. These bipyridine-centered triblocks were subsequently treated with $Fe(BF_4)_2.(H_2O)_6$ to form the iron(II) tris(bipyridine)-centered star-block copolymers (Scheme 84). The reactions took place in polar solvents, e.g. a mixture of dichloromethane and methanol. The chelation efficiency was affected by the triblock molecular weight and the composition.

4.2
Synthesis of Miktoarm Star (μ-Star) Copolymers

The term "miktoarm" (from the Greek word μικτός meaning mixed) is attributed to those star polymers with at least two different (in molecular weight, chemistry, or topology) blocks. So far, several strategies for the synthesis of μ-stars have been developed. These strategies involve the use of: multiheterofunctional initiators (MHI), multifunctional linking agents (MLA), divinyl compounds (DVC), 1,1-diphenylethylene derivatives, metal

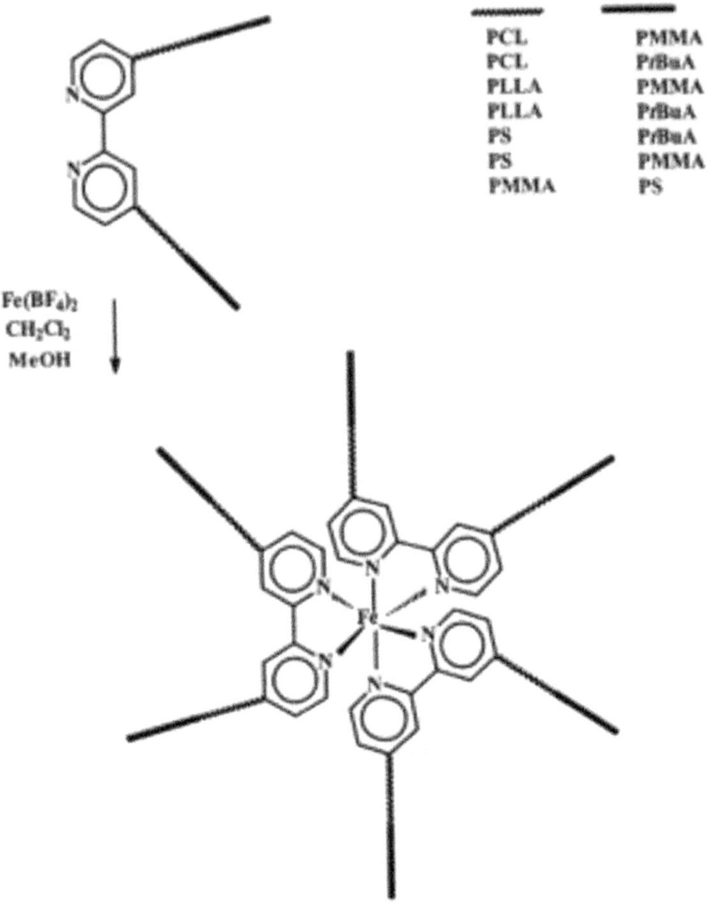

Scheme 84

templates or a combination of different polymerization techniques. Some of the above methods have been reviewed extensively (MLA, DVC), while others have received attention only in recent years. Our aim is to focus primarily on those examples given since 2000.

4.2.1 Multiheterofunctional Initiators

MHI possess at least two different polymerization initiating sites. The identical sites are selective for a particular class of monomers, and thus the resulting μ-star consists of chemically different arms. In order to obtain well-defined μ-stars, these identical active sites should have equal reactivity and furthermore, initiation should be faster than propagation. It is not always possible to achieve these requirements since differentiation in the topology of initiating sites causes differentiation in the polymerization rate. In addition, the accurate characterization of these species is extremely difficult.

The synthesis of the (PS)(PMMA)(PCL) 3μ-ABC star has been achieved with the use of a triheterofunctional initiator (ROP, ATRP, NMP) as shown in Scheme 85 [160]. SEC characterization experiments of the different intermediates confirmed the successful synthetic procedure.

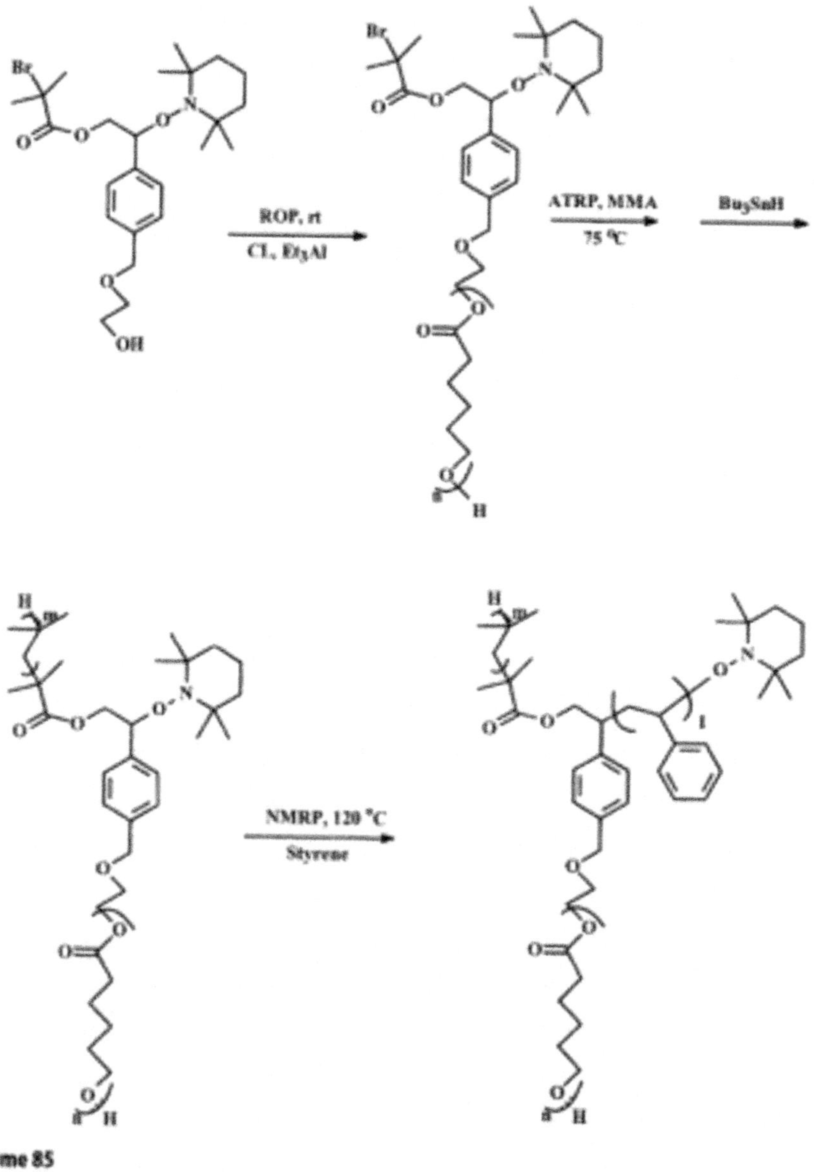

Scheme 85

Miktoarm stars of the $A(BC)_2$ type, where A is PS, B is poly(*t*-butyl acrylate) (PtBA), and C is PMMA [161] have been synthesized, by using the trifunctional initiator 2-phenyl-2-[(2,2,6,6-tetramethyl)-1-piperidinyloxy] ethyl 2,2-bis[methyl(2-bromopropionato)] propionate (NMP, ATRP) (Scheme 86). In the first step, a PS macroinitiator with dual *ω*-bromo functionality was obtained by NMP of styrene in bulk at 125 °C. This precursor was subsequently used as the macroinitiator for the ATRP of *tert*-butyl acry-

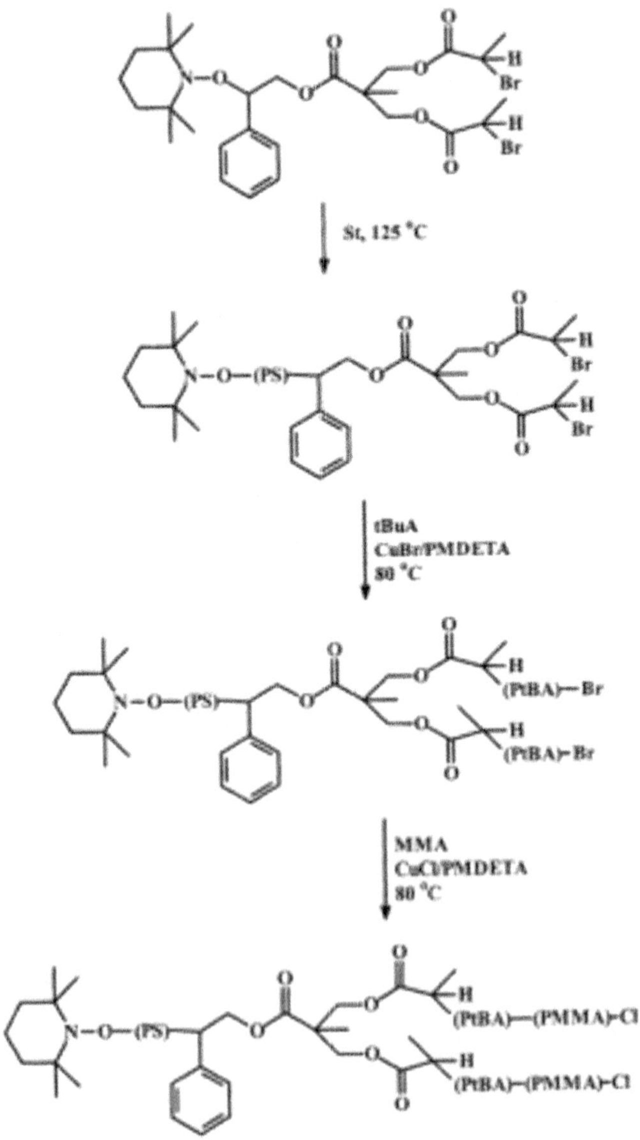

Scheme 86

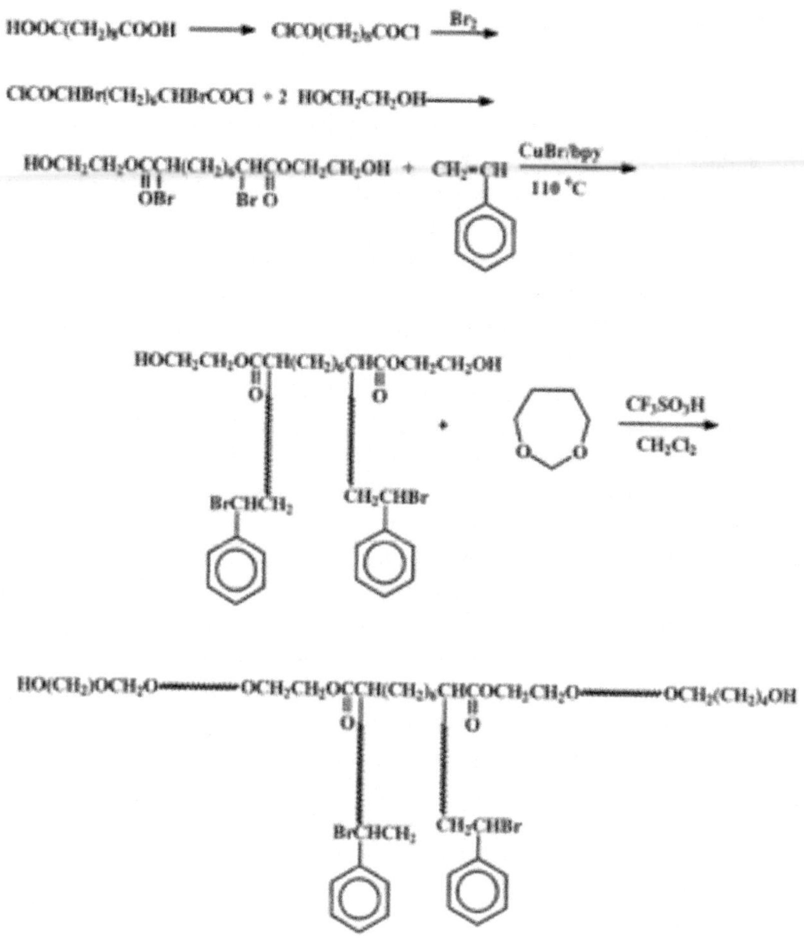

Scheme 87

late in the presence of CuBr and pentamethyldiethylenetriamine at 80 °C, to give the miktoarm star (PS)(PtBA)$_2$. This star was the macroinitiator for the polymerization of MMA, to afford the (PS)(PtBA-*b*-PMMA)$_2$ μ-stars.

A tetrafunctional initiator, was reported for the synthesis of A_2B_2 miktoarm stars, where A is PS and B is poly(1,3-dioxepane) (PDOP), respectively [162]. In the synthetic approach of Scheme 87, the bromide sites of di(hydroxyethyl)-2,9-dibromosebacate are the functional ATRP initiators for styrene. The remaining two hydroxyl groups serve as the initiating sites for the cationic ring opening polymerization of DOP in the presence of triflic acid. NMR and SEC characterization indicated the high degree of molecular and compositional homogeneity of the (PS)$_2$(PDOP)$_2$ stars.

4.2.2
Multifunctional Linking Agents

With the multifunctional linking strategy either homo- or heterofunctional linking agents can be employed. The homo-approach was widely used in the past for the synthesis of well-defined μ-stars with predetermined number and molecular characteristics of arms [163]. New applications of these approaches have been recently reported for the synthesis of A_2B, A_3B and A_2B_2 type μ-stars.

The synthesis of well-defined 3- and 4- miktoarm star copolymers of the A_2B and A_3B types, where A is 1,4-polybutadiene and B is poly(1,3-cyclohexadiene), has been carried out by using anionic polymerization and controlled chlorosilane chemistry [164]. Poly(1,3-cyclohexadienyl)lithium reacts with an excess of methyltrichlorosilane or tetrachlorosilane followed, after the elimination of the excess silane, by the addition of a slight excess of polybutadienyllithium. Characterization by SEC, low angle laser light scattering, LALLS, laser refractometry and NMR spectroscopy reveal a high degree of molecular and compositional homogeneity. Heterogeneous catalytic hydrogenation of the polydiene μ-stars, leads to μ-stars containing one amorphous polycyclohexylene arm (high T_g) as well as either two or three crystalline polyethylene arms. Using the same methodology, well-defined $(PS)(P2MP)_2$ and $(PS)(P2MP)_3$ star copolymers have also been synthesized, where P2MP is poly(2-methyl-1,4-pentadiene) [165].

In order to probe the effect of junction point functionality on chain conformation and morphology of miktoarm star block copolymer architectures, a series of PI_nPS_n ($n = 2, 4, 16$) was synthesized [166]. A single batch of both living PS and PI arms have been used, in order to ensure that all chemically identical arms (either A or B) have the same molecular weights. The living A and B chains were reacted with the appropriate chlorosilane, under appropriate experimental conditions, to produce the corresponding μ-stars, as shown in Scheme 88.

The heterofunctional linking approach, in spite of its potential efficiency, has not been yet explored for the synthesis of μ-stars. A reaction procedure is given in Scheme 89. Living PDMS is selective only for Si – Cl groups. The

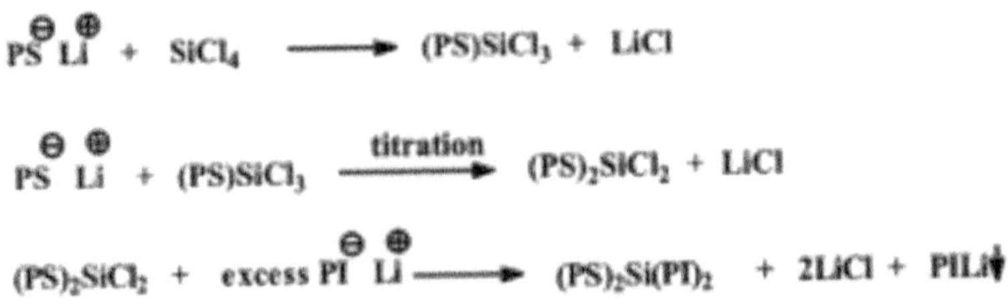

Scheme 88

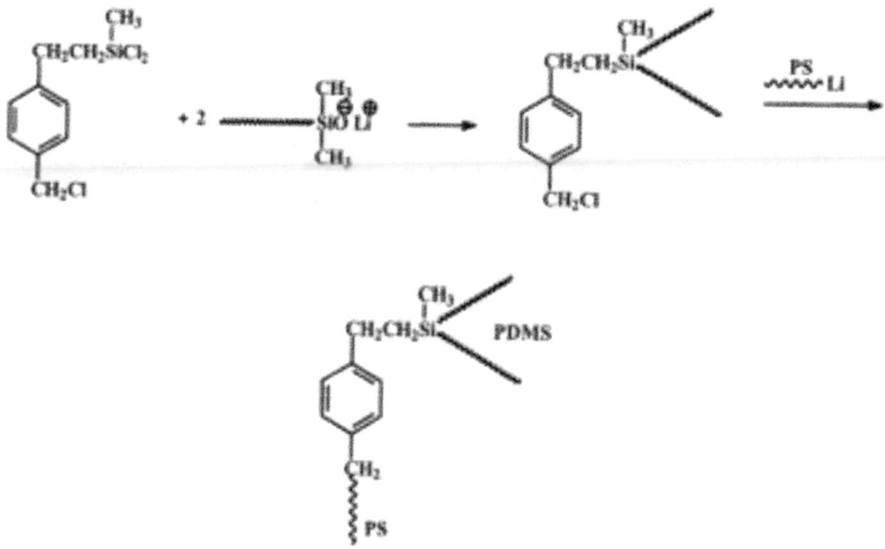

Scheme 89

remaining $-CH_2Cl$ group can react with several living chains i.e. PSLi, PILi, PBdLi, P2VPLi etc. to produce miktoarm stars.

4.2.3 Divinyl Compounds

These compounds can either be homopolymerizable (e.g. divinylbenzene, divinylethers) or non-homopolymerizable (e.g. double diphenylethylenes DDPE). The use of divinylbenzene, DVB, was first recognized by Eschwey and Burchard [167] and developed mainly by Rempp and colleagues [168–171]. The general reactions are given in Scheme 90.

The living ALi chains polymerize a small amount of DVB leading to the formation of a star molecule bearing within its core (microgel nodule of DVB) a number of active sites, which is theoretically equal to the number of incorporated A arms. Subsequent addition of monomer B yields the μ-star copolymer.

The double diphenylethylenes (DDPE) approach was first reported by Höcker and Latterman [172] and was later developed mainly by the Quirk (anionic) [173, 174] and Faust (cationic) groups [175, 176]. Representative reactions for the synthesis of the A_2B_2 4μ-star by the anionic route are shown in Scheme 91.

The divinyl compound approach has been extensively covered in previous reviews. However, an interesting example for the synthesis of A_2B_2 4μ-stars, where A is either PI or PBd and B is PMMA or PBMA, has recently been prepared [177]. Following polymerization of the diene in hexane by *s*-BuLi,

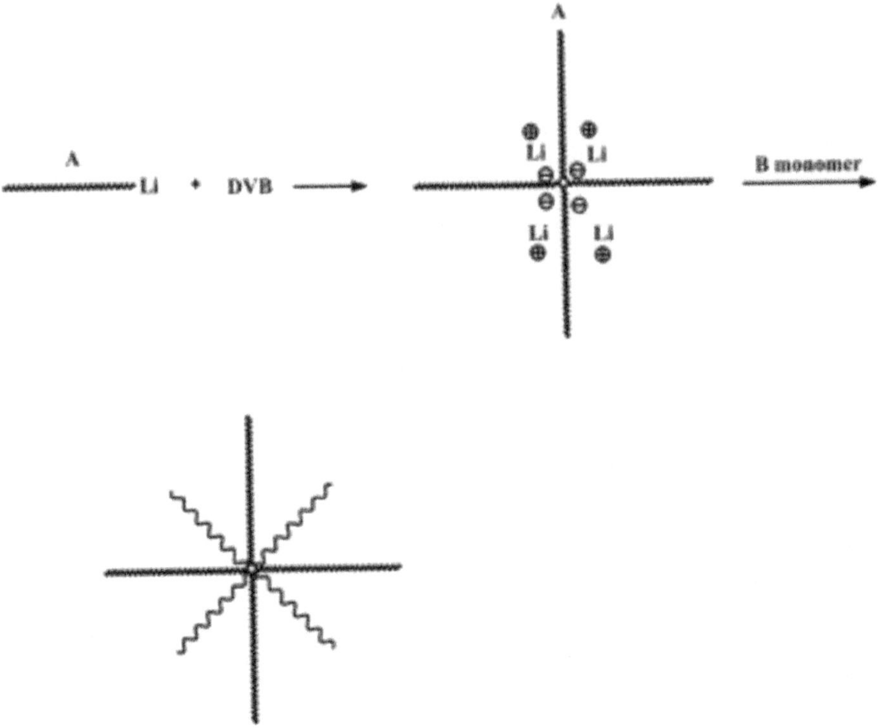

(μ-A_nB_n)

Scheme 90

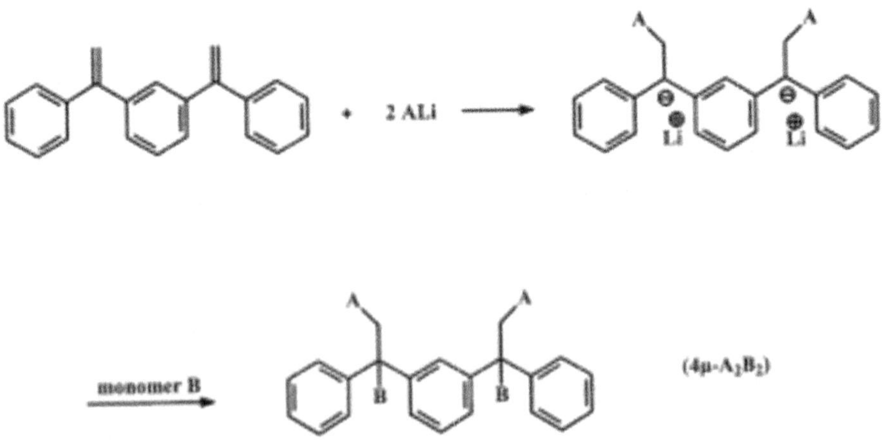

(4μ-A_2B_2)

Scheme 91

the solvent was changed to THF and the living polydiene chains were linked in pairs to 1,2-bis[4-(1-phenylethenyl)]ethane. The two new active sites generated were used for the polymerization of the methacrylate monomers at – 78 °C. Extensive characterization affirmed the structure claimed.

4.2.4 Diphenylethylenes (DPE)

This strategy, based on 1,1-diphenylethylenes, which are non-homopolymerizable monomers, has been developed mainly by the Quirk [174] and Hirao [178] groups. DPEs continue to find applications for the synthesis of the μ-stars. A few recent examples are given below.

Three different ABC 3μ-stars, where A is always PS, B is either PEO or PMMA, and C is poly(ε-caprolactone), poly(L-lactide) or PEO have been synthesized by similar procedures [179] (Scheme 92). Living arm A, was obtained by using cumyl potassium as the initiator, and was subsequently reacted with the double bond of (1-[4-(2-*tert*-butyldimethylsiloxy)ethyl]phenyl-1-phenylene) to give a living end functionalized polymer with a protected – OH group. The active anions were used for the polymerization of EO, leading to the formation of the second arm. After deprotection, the – OH group was transformed with diphenylmethyl potassium to the corresponding potassium alcholate, which acted as the initiating sites for the polymerization of either the ε-caprolactone or the L-lactide. For the synthesis of (PS)(PMMA)(PEO) miktoarm star, the same synthetic procedure was followed, with MMA and EO instead of EO and ε-caprolactone. The molecular characterization indicated relatively low polydispersity indices (~ 1.2) which implies a high degree of molecular and compositional homogeneity.

By using anionic polymerization techniques and (1-[4-(2-*tert*-butyldimethylsiloxy)ethyl]phenyl-1-phenylene), μ-stars of A_2B and A_3B type, where A is PS and B either PEO or PtBuMA were synthesized [180]. As an example the reaction sequence for the synthesis of $(PS)_2(PtBuMA)$ is given in Scheme 93. Depending on the polarity of the medium the reaction between PSLi and the DPE leads to incorporation of two or three PS chains by a nucleophilic substitution reaction at the benzylic carbon atom of the DPE unit. The new anionic site created is used for the polymerization of either EO or *t*BuMA. In the case of the EO, a phosphazine base was used in order to increase the reactivity of the anion and successfully polymerize the monomer. The high molecular weights obtained along with the low polydispersity and the good agreement between the stoichiometric and experimental molecular weights of the copolymers, indicate the high degree of molecular and compositional homogeneity.

In an extension of the methodology involving DPEs, the preparation of chain-end and in-chain functionalized polymers with a definite number of chloromethylphenyl or bromomethylphenyl groups and their utilization in

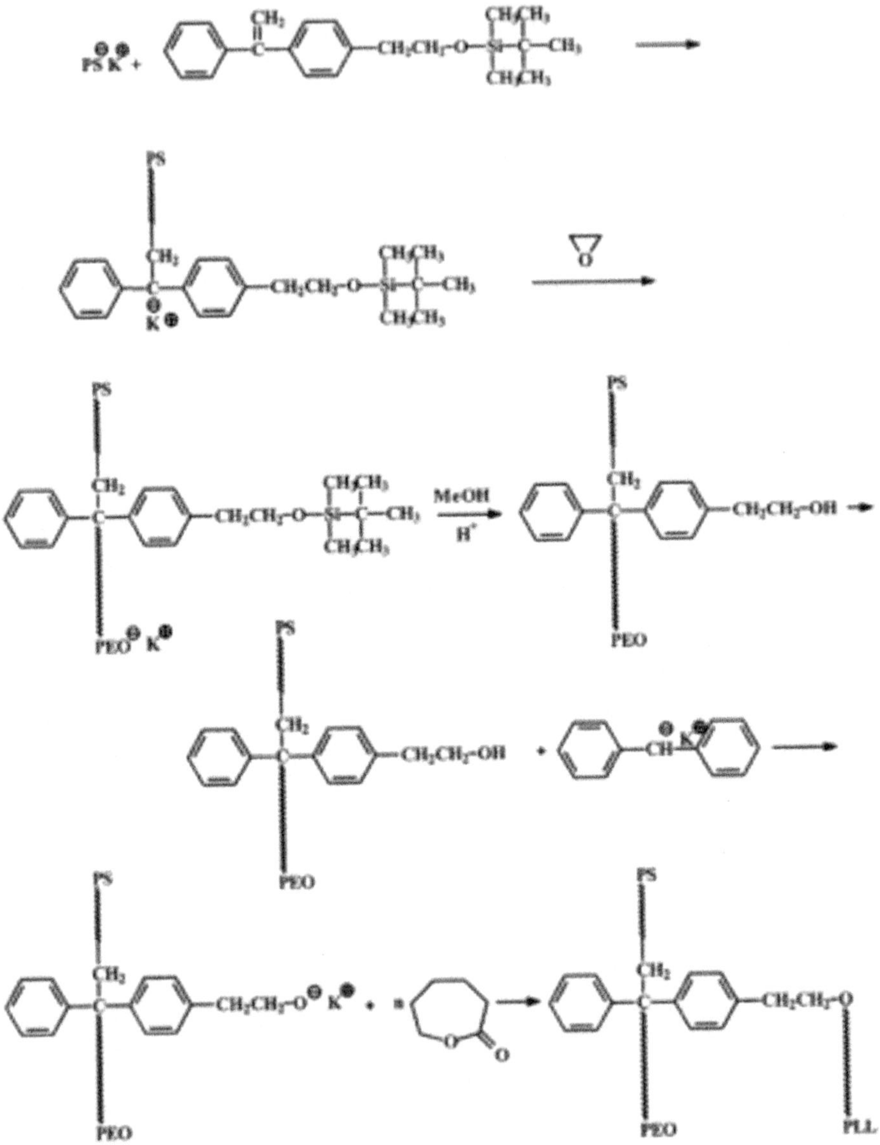

Scheme 92

the synthesis of miktoarm star polymers has been reported [178]. A macroanion is reacted with a DPE derivative having two methoxymethyl groups at the meta-positions of the phenyl rings (Scheme 94). After deactivation with methanol, the methoxymethyl groups can be converted quantitatively to chroromethyl phenyl groups (CMP) by reaction with BCl_3. These CMP groups are linking sites to other living polymeric chains. Using CMP-functionalized polystyrenes, along with appropriate DPE derivatives, a variety of miktoarm

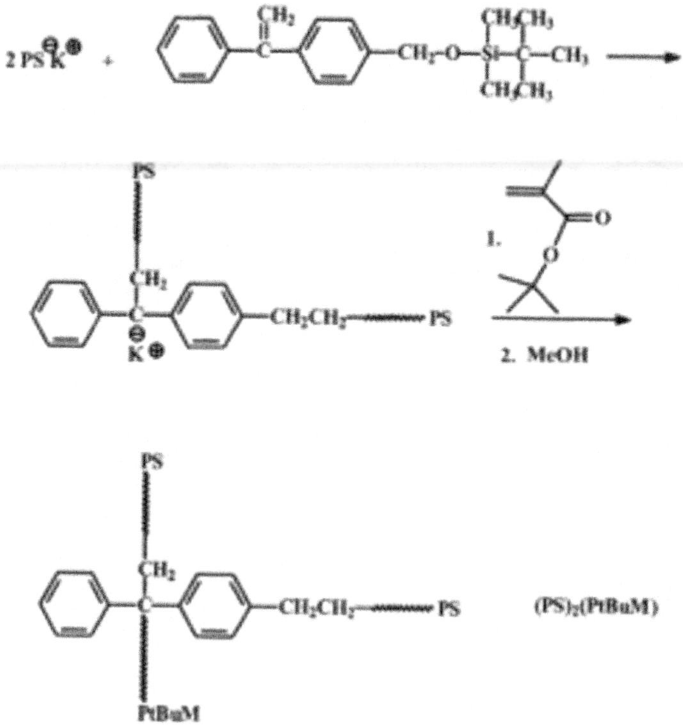

Scheme 93

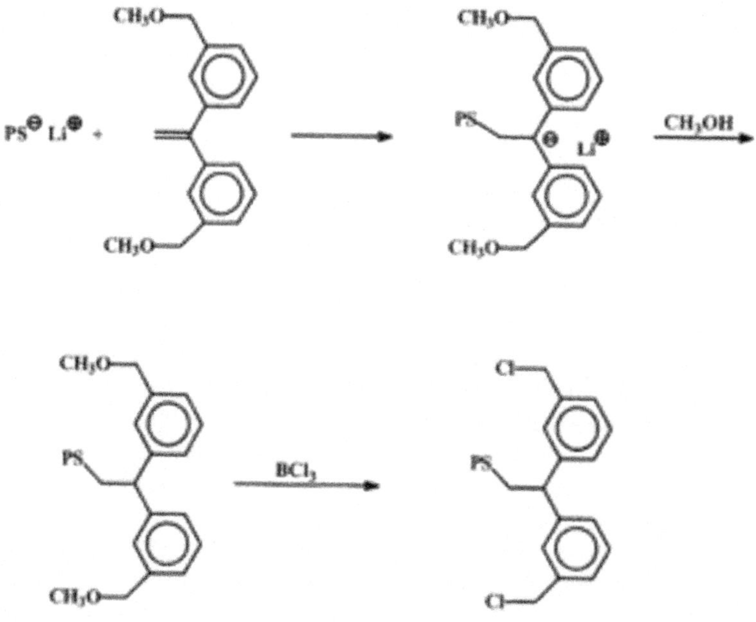

Scheme 94

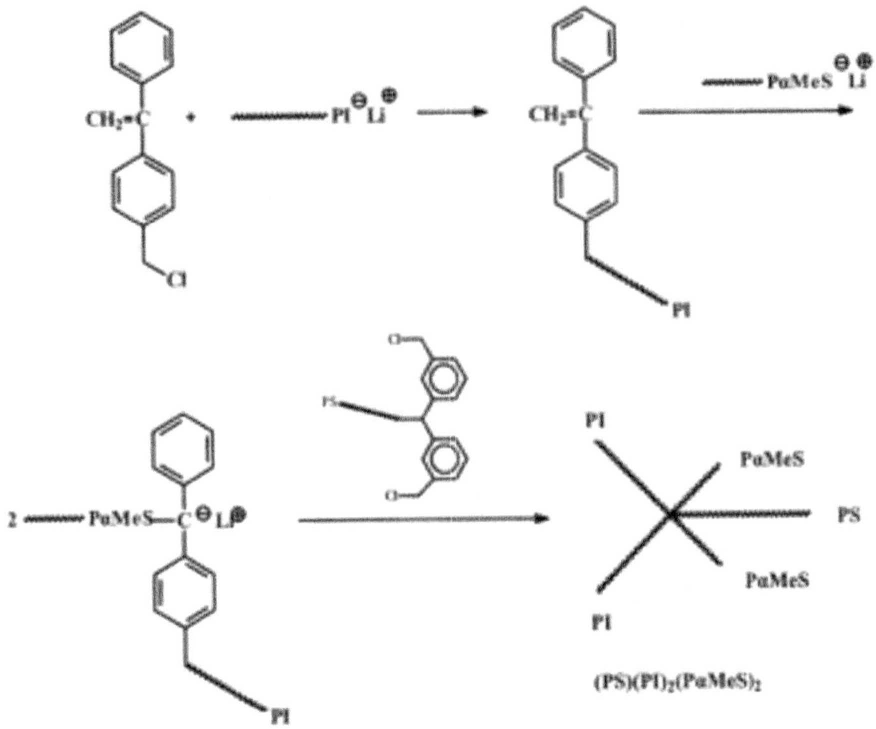

Scheme 95

star terpolymers of the ABC_2, ABC_4 and AB_2C_2 were synthesized, where A, B and C are PS, PI and poly(α-methylstyrene)(PαMeS), respectively. As an example, the reactions used for the synthesis of one of the most complex star architectures, $(PS)(PI)_2(P\alpha MeS)_2$, are given in Scheme 96.

4.2.5 Metal Template-Assisted Synthesis

Very recently a promising new method, based on metal complexes, has been reported for the synthesis of μ-stars. To our knowledge only the Fraser group has employed Ru to link ω-bipyridyl (bpy) PS and PMMA chains in the synthesis of well-defined $(PS)_2(PMMA)$ and $(PS)_4(PMMA)_2$ [181]. The PS and PMMA macroligands were synthesized by ATRP with initiators containing the bpy group (Scheme 96). By chelation of the resulting bpyPS macroligands, under conditions where only two bpy groups could be attached on each Ru atom, complexes of the $Ru(PS)_{2n}$ [Scheme 96, (1)] type were formed. These complexes were subsequently reacted with one bpyPMMA to produce $(PS)_2(PMMA)$ [Scheme 96, (2)]. In the case of $(PS)_4(PMMA)_2$, a difunctional initiator containing two bpy groups at each chain end was used (Scheme 96).

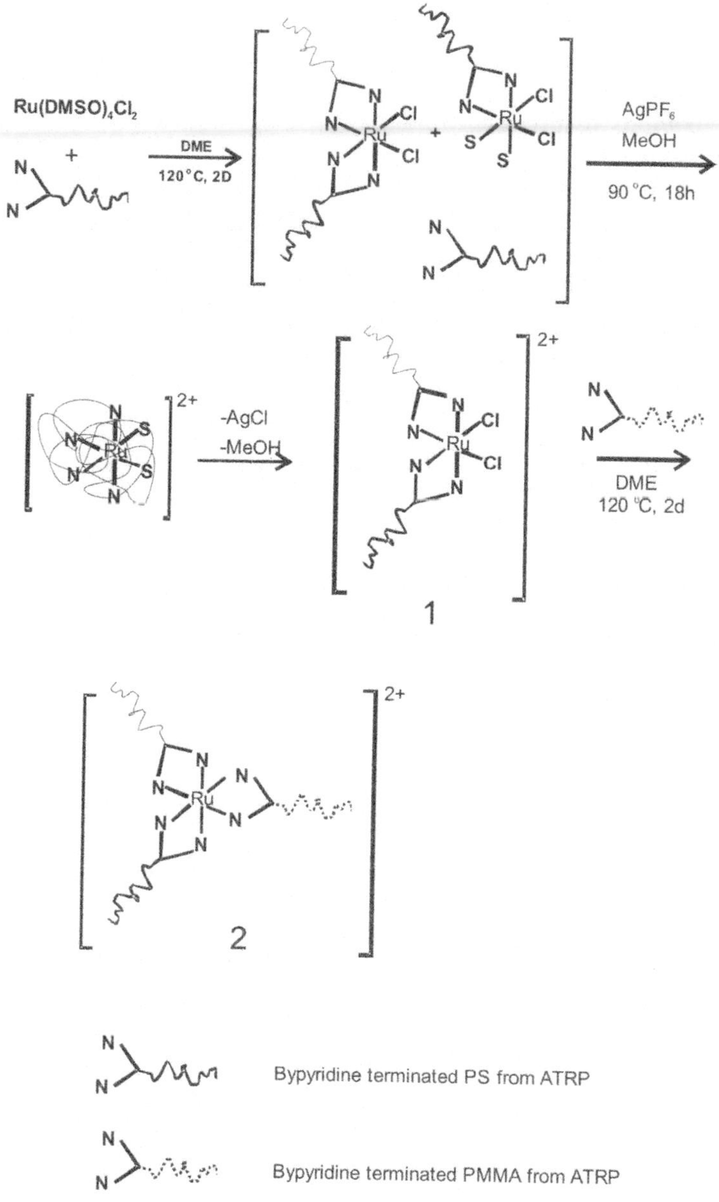
Ru(DMSO)$_4$Cl$_2$
+
DME
120 °C, 2D
AgPF$_6$
MeOH
90 °C, 18h
2+
-AgCl
-MeOH
DME
120 °C, 2d
1
2
Bypyridine terminated PS from ATRP
Bypyridine terminated PMMA from ATRP
S: solvent molecule

Scheme 96

4.2.6 Combinations of Polymerization Techniques

This category comprises those methods which combine different polymerization methods to produce μ-stars. The initiating sites for the different polymerizations are created, step-by-step, during the μ-star synthesis.

By using a combination of RAFT and ring opening polymerization (ROP), (poly(ethylene oxide) methyl ether)(polystyrene)(poly(L-lactide) 3-miktoarm star terpolymers have been successfully synthesized [182]. The synthetic approach involved the reaction of the ω-functionalized –OH group of the poly(ethylene oxide) methyl ether with maleic anhydride under conditions where only one hydroxyl group can be esterified (MPEO). The double bond

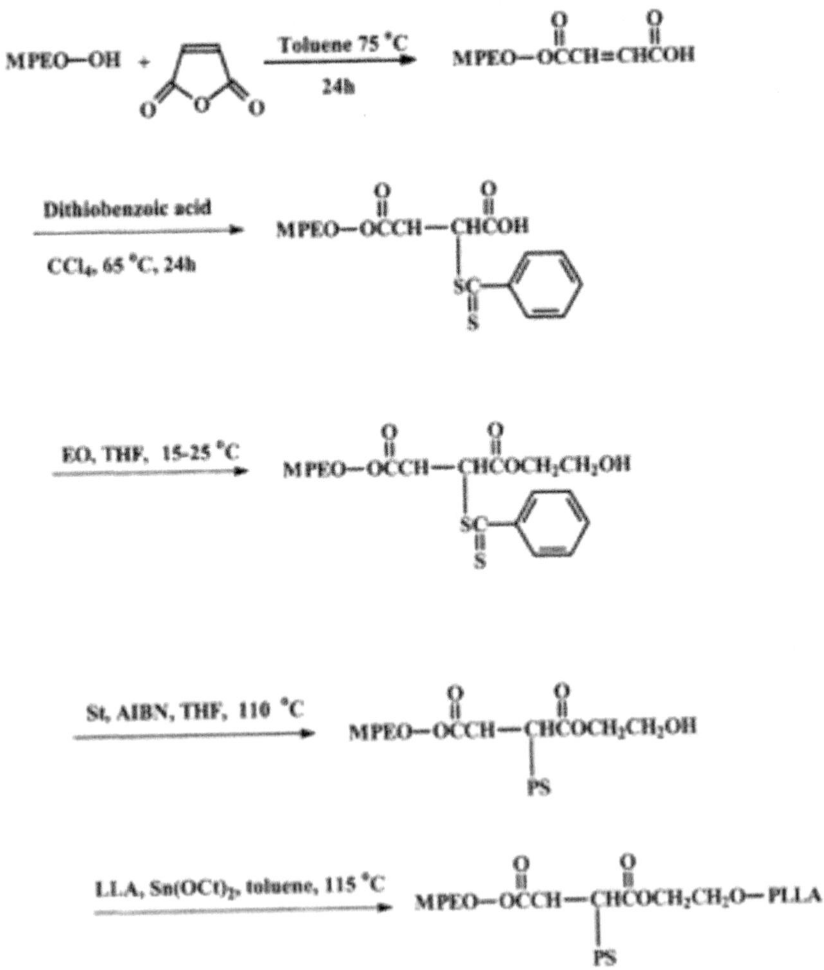

Scheme 97

of the maleic group was then reacted with dithiobenzoic acid to afford dithiobenzoic-terminated MPEO. The second carboxyl group of the maleic anhydride was then reacted with ethylene oxide to give the corresponding ester with an – OH group. The dithiobenzoic group of the MPEO was used for the RAFT polymerization of styrene and the OH group for the ROP of L-lactide (Scheme 97). The intermediate products along with the final terpolymers were characterized by SEC and NMR spectroscopy. The polydispersity indices of the intermediate along with the final products were between 1.05–1.07, indicating the high degree of molecular and compositional homogeneity.

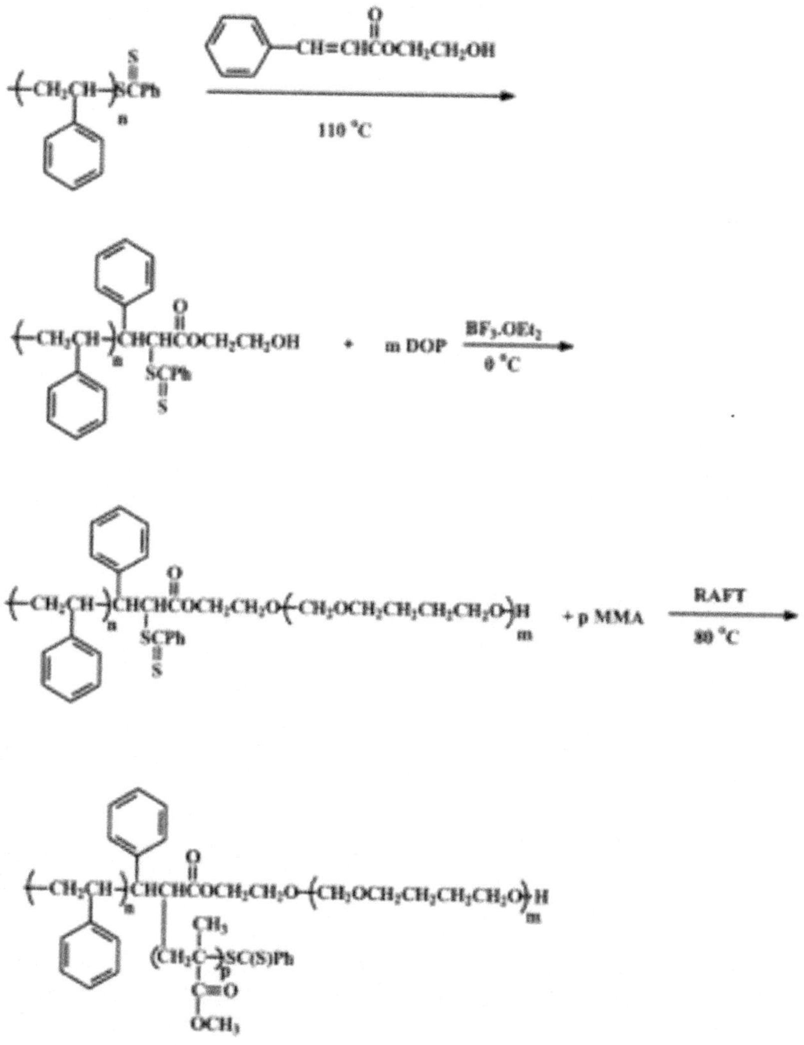

Scheme 98

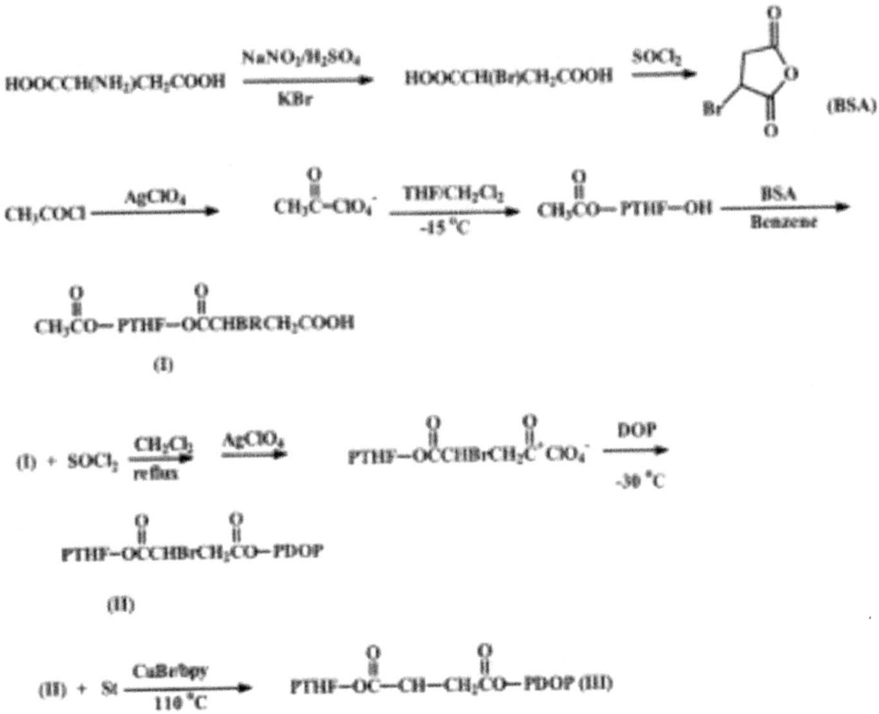

Scheme 99

By utilizing a combination of RAFT and cationic ROP, the synthesis of [poly(methyl methacrylate)][poly(1,3-dioxepane)][polystyrene] miktoarm star terpolymers was achieved [182]. The approach involved the synthesis of PS functionalized with a dithiobenzoate group by RAFT polymerization and subsequent reaction with hydroxyethylene cinnamate (Scheme 98). The newly created hydroxyl group was then used for the cationic ring opening polymerization of 1,3-dioxepane (DOP). The remaining dithiobenzoate group was used for the RAFT polymerization of methyl methacrylate.

A third example combines cationic ROP and ATRP for the synthesis of (polytetrahydrofurane)(poly-1,3-dioxepane)(PS) miktoarm stars (Scheme 99). The initiating sites for the above polymerization were created step-by-step from amino-succinic acid (Scheme 99).

4.3 Synthesis of Graft Copolymers

Graft copolymers are comb-shaped polymers consisting of a backbone and two or more branches which differ chemically from the backbone. Branches are usually distributed randomly along the backbone, although recent advances in synthetic methods have allowed for the preparation of better-defined structures. Comb-shaped polymers can be prepared by three general synthetic methods: "grafting onto", "grafting from", and "grafting through", shown schematically in Scheme 100.

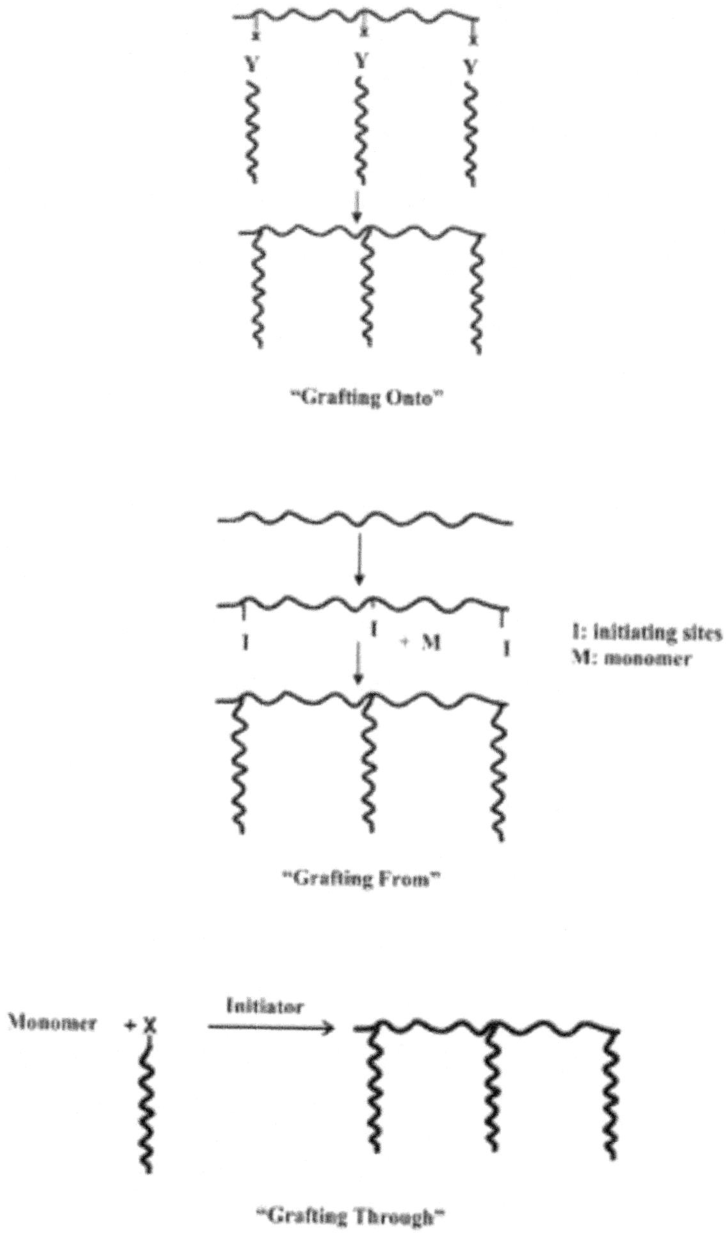

Scheme 100

In the "grafting onto" method the backbone and the arms are prepared separately by a living polymerization mechanism. The branching sites can be introduced onto the backbone either by post polymerization reactions or by copolymerization of the main backbone monomer(s) with a suitable comonomer, with the desired functional group (unprotected or in a protected form if this functional group interferes with the polymerization reaction). The average number of branches can be estimated from the molecular weight of the final graft copolymer and the known molecular weight of the backbone and the branches. The "grafting onto" synthesis of poly(styrene-g-ferrocenyldimethylsilane) has been recently reported [183]. The formation of the backbone involved copolymerization of styrene and chloromethylstyrene by conventional radical initiators. Subsequently, the chloromethyl groups of the backbone reacted with living polyferrocenyldimethylsilane, previously synthesized by anionic ROP of sila [1–3] ferrocenophane with *s*-BuLi as the initiator (Scheme 101).

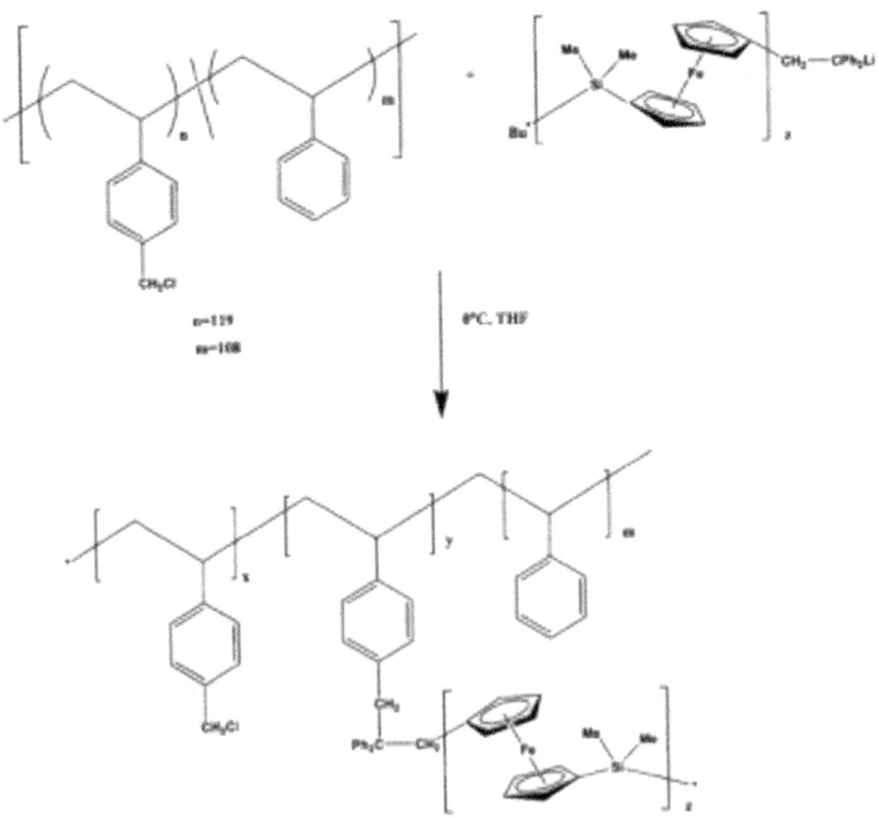

Scheme 101

Metallo-supramolecular graft copolymers poly(MMA-*g*-PEO) and poly(MMA-*g*-L-lactide) were also synthesized by the "grafting onto" strategy [184]. Conventional radical copolymerization of MMA with methacrylate-modified monomers, containing terpyridine functional moieties resulted in the formation of a PMMA backbone with terpyridine units along the polymeric chain. Either ω-functionalized PEO or poly(L-lactide) terpyridine/ruthenium(III) mono-complexes (7) or (8) were then grafted onto the terpyridine groups of PMMA through the double-complexes (Scheme 102). The polydispersity index of the copolymers were as low as 1.2. However, the molecular weights of the final copolymers were rather low, and the average number of grafted chains was about 2.

In the "grafting from" method, after the preparation of the backbone, active sites are produced along the backbone that are able to initiate the polymerization of a second monomer(s), thus forming the branches. The number of branches can be controlled by controlling the number of active sites generated along the backbone assuming that each one of them could initiate the polymerization. Obviously, the isolation and characterization of each part of the graft copolymer in this case is extremely difficult.

A recent example is provided by Liu and Sen [185], who prepared the poly(ethene-*co*-styrene) backbone (having a polydispersity index equal to 2.7) by copolymerization of ethylene and styrene using $[C_5Me_4(SiMe_2NtBu)]TiCl_2$/MAO as the catalyst. In order to create the initiating sites along the backbone, the labile tertiary hydrogen atom of the styrenic units were substituted by bromine groups (Scheme 103). The *tert*-Br groups subsequently served as the initiating sites for the ATRP of MMA, in the presence of CuBr and PMDETA to give poly(ethene-*co*-styrene)-*g*-poly(methyl methacrylate). By using the same strategy, poly(ethene-*co*-styrene)-*g*-polystyrene, poly(ethene-*co*-styrene)-*g*-(poly(methyl methacrylate-*b*-polystyrene)) and poly(ethene-*co*-styrene)-*g*-(poly(methyl methacrylate)-*b*-poly(2-hydroxyethyl methacrylate)) were also synthesized.

The same group employed the "grafting from" method to synthesize poly(β-pinene)-*g*-polystyrene [186]. The poly(β-pinene) backbone was synthesized by living cationic polymerization with the 1-phenylethyl chloride/$TiCl_4$/$Ti(OiPr)_4$/nBuNCl initiating system. Bromination of the poly(β-pinene) leads to the formation of $-$CHBr groups along the backbone which are the initiating sites for the ATRP of styrene (Scheme 103). The copolymers, characterized by SEC and NMR spectroscopy exhibited rather broad molecular weight distributions ($\sim$ 1.5).

The "grafting from" methodology was also utilized for the synthesis of poly(4-methylphenoxyphosphazene-*g*-2-methyl-2-oxazoline) graft copolymers [187]. The synthetic approach involved the thermal polymerization of hexachlorophosphazene, in the presence of aluminum chloride, to give low molecular weight poly(dichlorophosphazene). The chloro groups were subsequently replaced by 4-methylphenoxy groups, followed by partial bromi-

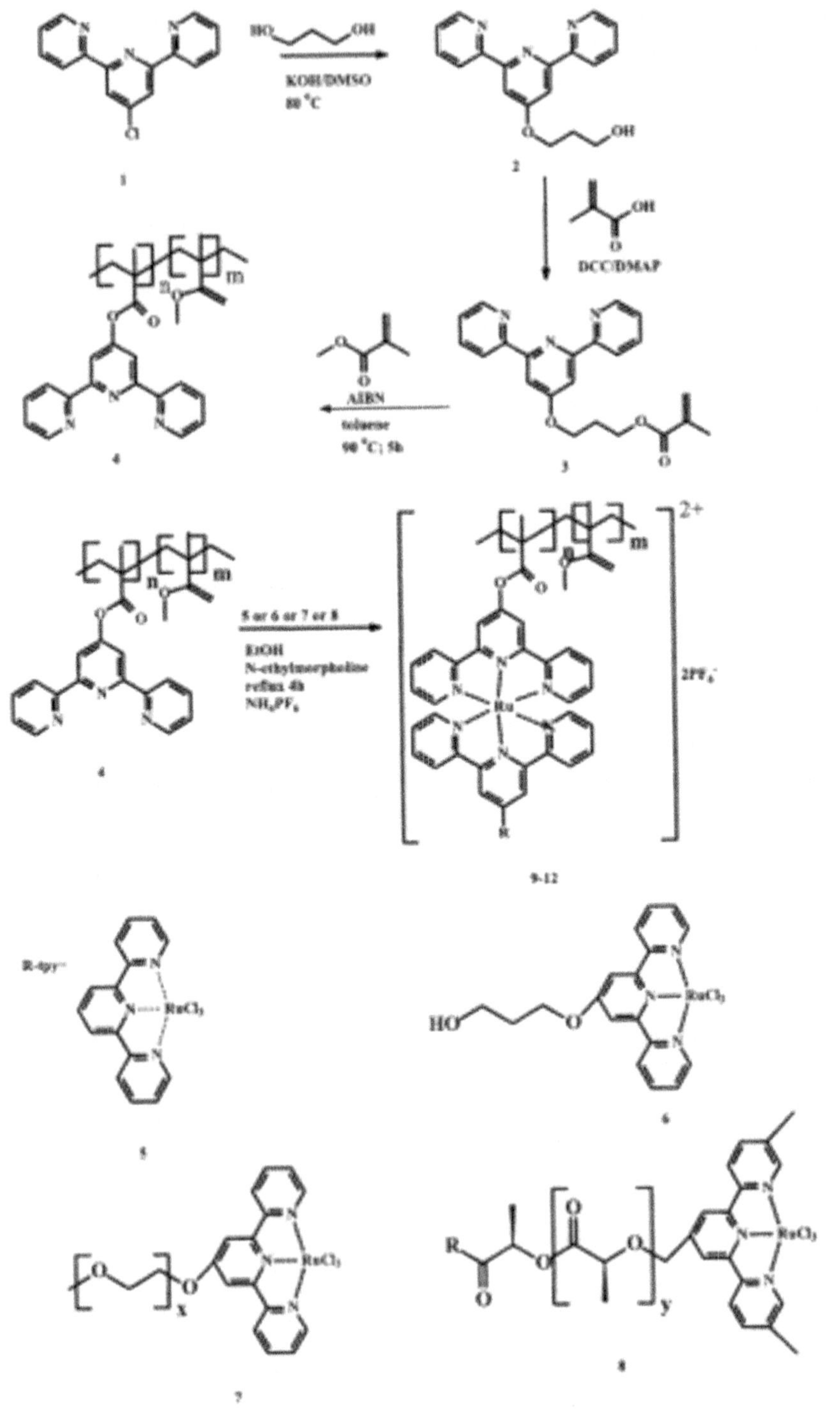

Scheme 102

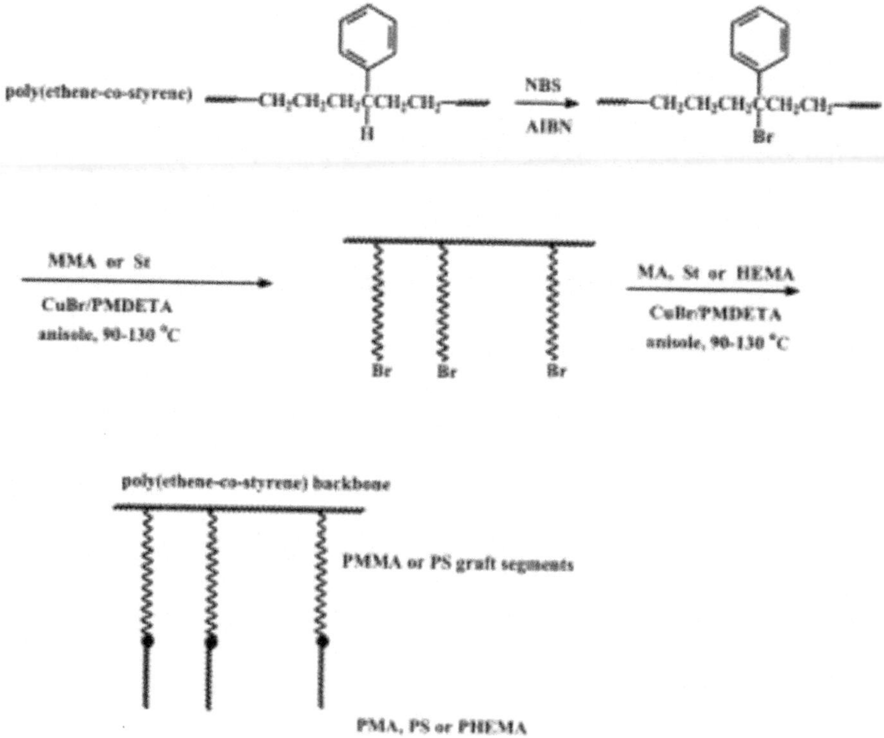

Scheme 103

nation of the methyl groups. The resulting brominated polymer was the macroinitiator for the cationic ROP of 2-methyl-2-oxazoline at 80 °C in DMF.

In the "grafting through" or macromonomer method, preformed macromonomers are copolymerized with a conventional monomer in order to produce the graft copolymer. In this case the macromonomer side chains are the branches of the final graft copolymer with the backbone formed *in situ*. The number of branches per backbone can be generally controlled by the ratio of the molar concentrations of the macromonomer and the comonomer. However, several other factors have to be considered, the most important being the reactivity ratios of the macromonomer and the comonomer, since these ratios can influence the placement of the branches along the backbone (tapered, random, blocky). The "grafting through" graft copolymers appear to be the most favorable and promising category since no fractionation is needed to isolate the graft copolymers from either backbone and/or branches. In addition, if appropriate conditions are applied, this strategy has the potential to lead to a wide variety of well-defined structures. The importance of the "grafting through" methodology is clear from the many examples which currently appear in the literature.

The synthesis of polystyrene-g-polytetrahydrofurane [188] was achieved by ATR copolymerization of methacrylic PTHF macromonomer, MA-PTHF, with styrene (Scheme 105). The PTHF macromonomer was synthesized by cationic ring opening polymerization of THF with acrylate ions, formed by the reaction of methacryloyl chloride and $AgClO_4$. The polydispersity indices of the graft copolymers determined by SEC ranged between 1.3–1.4. Kinetic studies revealed that the relative reactivity ratio of the macromonomer to St was independent of the molecular weight of PTHF.

By using the same strategy, poly(propene-g-styrene) graft copolymers were prepared [189]. Allyl-terminated polystyrenes (PS macromonomers) were synthesized by ATRP of styrene followed by carbocationic chain end transformation with allyltrimethylsilane in the presence of titanium tetrachloride. Systematic investigations were performed in order to examine the influence of the molecular weight, the type of catalyst, the polymerization temperature, and the propene pressure on the metallocene/MAO-catalyzed copolymerization of the PS macromonomers with propene. The resulting materials were characterized by SEC and NMR. Relatively high polydispersity indices (~ 1.7–2.1) were found for the graft copolymers.

Polystyrene-g-poly(ethylene oxide) was synthesized by the copolymerization of styrene and styrenic PEO with $CpTiCl_3$/MAO catalyst [190]. In this case the macromonomer was prepared by first reacting the sodium salt of PEO-OH with NaH and then with a 5-fold amount of *p*-chloromethyl styrene.

Thiol end-functionalized poly(2-hydroxyethyl methacrylate-g-ethylene glycol) graft copolymers were synthesized by ATRP copolymerization of

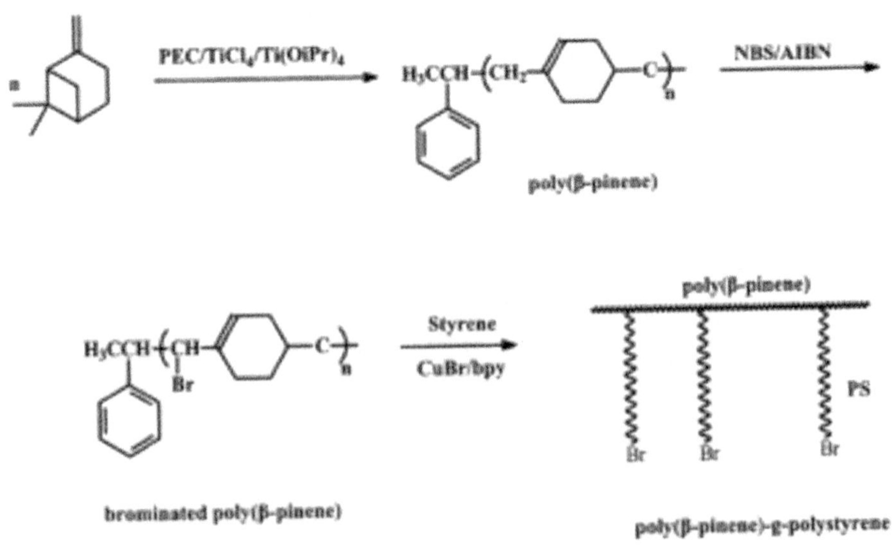

Scheme 104

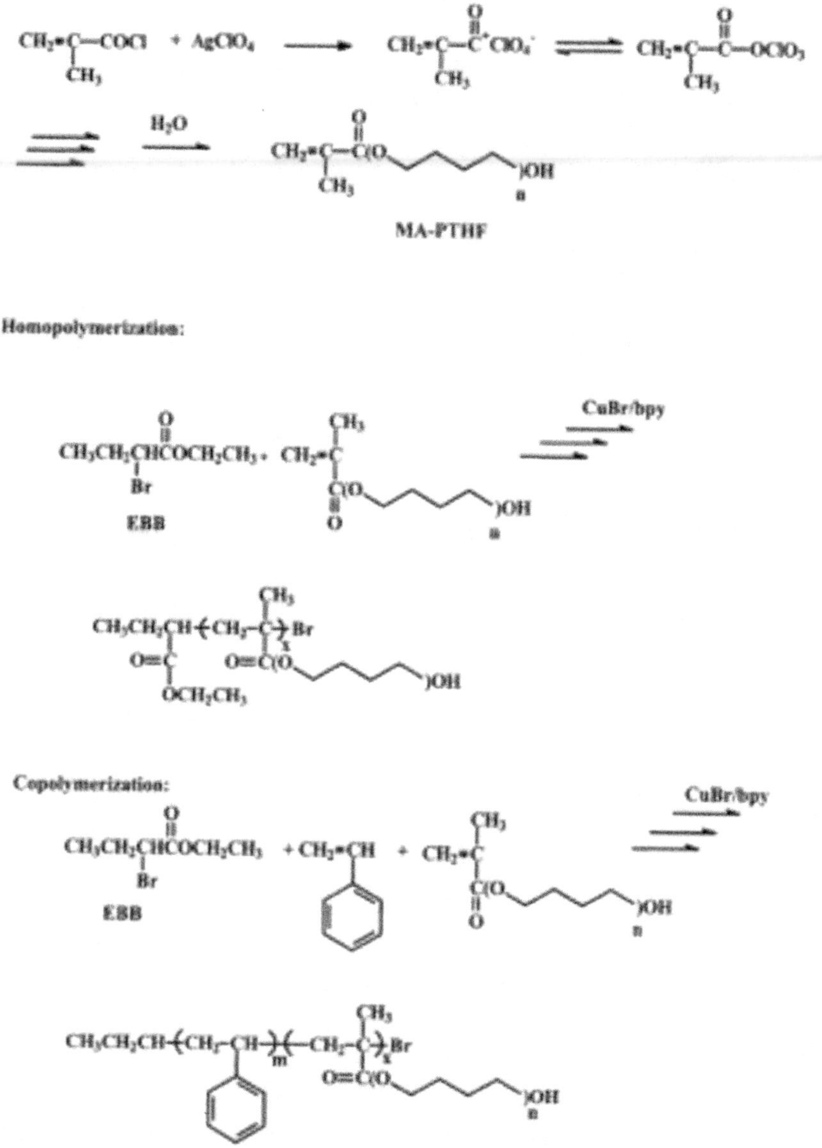

Scheme 105

2-hydroxyl methacrylate monomer and PEO macromonomers containing a methacrylate group at one end [191]. The initiator used was 2-(2,4-dinitrophenylthio)ethyl 2-bromo-2-methylpropionate. After the completion of the polymerization, the 2-(2,4-dinitrophenylthio)ethyl group was replaced by the 2-mercaptoethanol group, by transesterification according to Scheme 106. The copolymers were extensively characterized by NMR, and SEC-MALLS

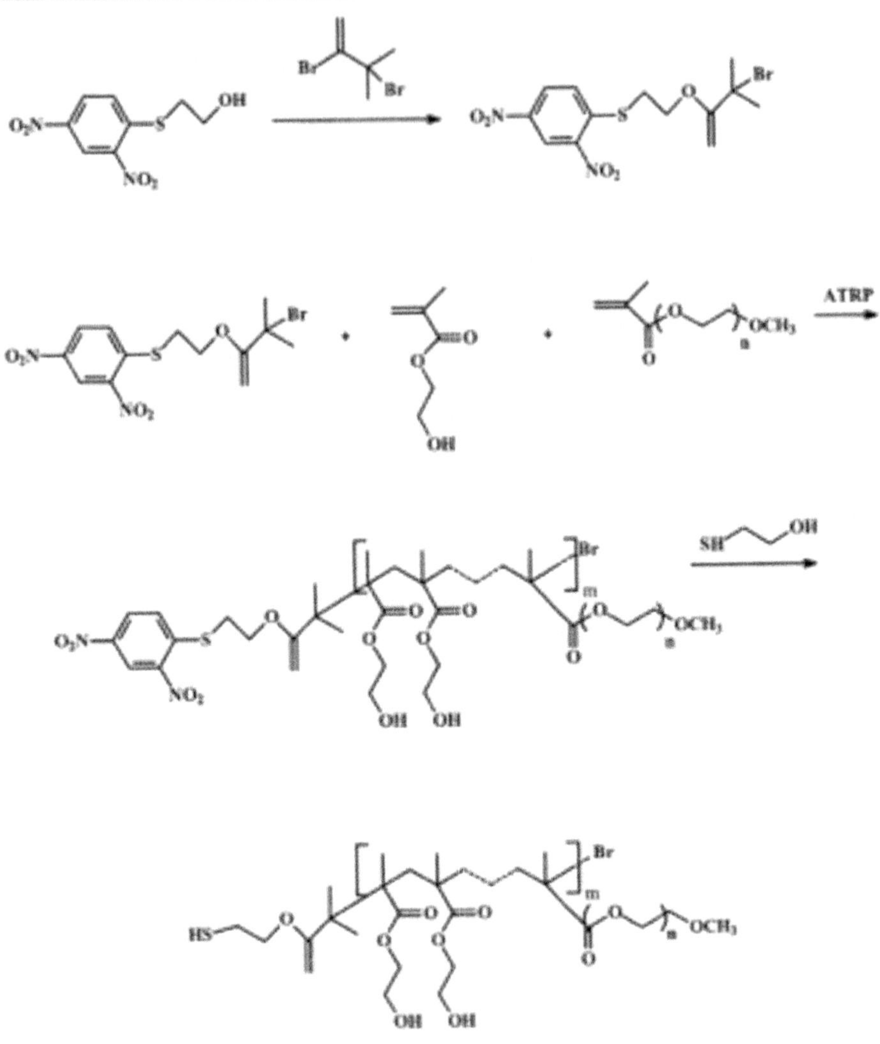

Scheme 106

chromatography. The thiol-terminated graft copolymers were attached on gold-coated silicon wafers, and their conformation was examined by AFM.

An alternative route for the preparation of styrenic macromonomers is the reaction of living chains with 4-(chlorodimethylsilyl)styrene (CDMSS) [192]. The key parameter for the successful synthesis of the macromonomers is the faster reaction of the living anionic chain with the chlorosilane group rather than with the double bond of the CDMSS. Anionic *in situ* copolymerization of the above macromonomers (without isolation) with conventional monomers leads, under appropriate conditions, to well-defined comb-like chains with a variety of structures.

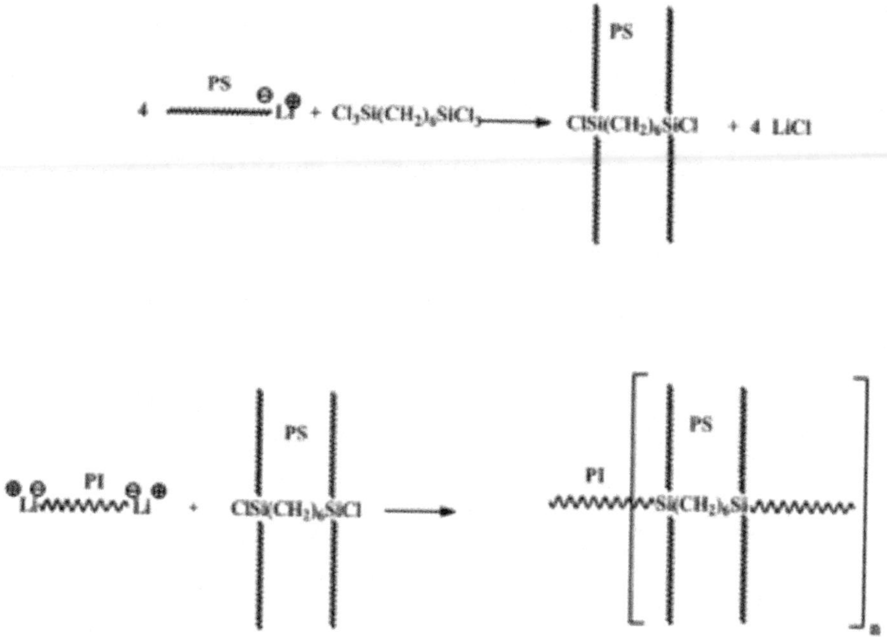

Scheme 107

Finally, examples are reported in which two grafting methods ("onto", "through") are combined to produce multigraft poly(isoprene-g-styrene) copolymers with tri-, tetra- and hexafunctional branched points [193, 194]. The synthetic strategy employs classic anionic polymerization techniques and utilizes a modular approach in which polystyryllithium and α,ω-poly(1,4)isoprenyldilithium are sequentially added into chlorosilane linking centers with different functionalities. Living PSLi is added either to trichlorosilane, tetrachlorosilane or 1,6-bis-(trichlorosilyl)hexane, in an incremental way, until one, two, or four equivalents have been incorporated into the linking agent, respectively. Then, a difunctional LiPILi was added in a slight excess, resulting in condensation between macromolecular dinucleophiles and dielectrophiles. The reactions used in the case of the regular grafts with the hexafunctional branched points are shown in Scheme 107. The copolymers resulting from the trifunctional linking agent were combs, while those synthesized by the terafunctional and hexafunctional linking agents were named "centipedes" and "barbwires".

4.4 Synthesis of Cyclic Copolymers

Living polymerization processes leading to linear copolymeric precursors, with either identical or complementary functional end groups (X,Y) have been used for the synthesis of cyclic block copolymers. In the first case, an α,ω-homodifunctional copolymeric chain reacts with an appropriate difunctional linking agent as shown below.

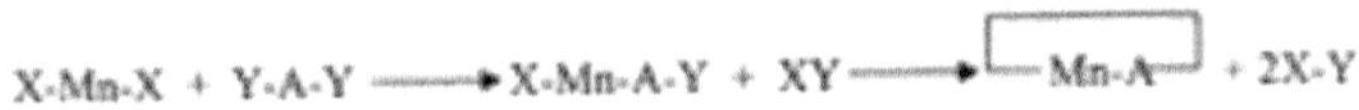

Besides the expected intramolecular reaction, several other side reactions can occur:

These reactions lead to undesirable high molecular weight polycondensates, which are either linear or cyclic, and should be removed from the desirable low molecular weight cyclic product.

In the second case the intramolecular cyclization requires a catalytic activation step:

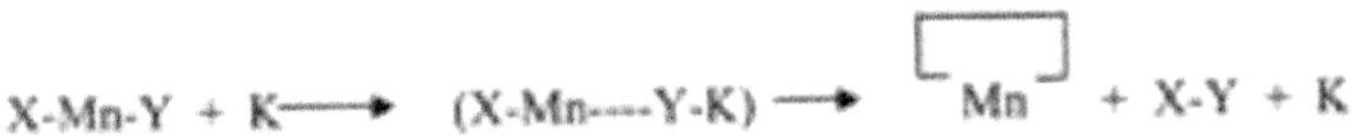

This synthetic approach presents several advantages, namely that the exact stoichiometry of the two reagents is not required, since the two reactive groups are in the same molecule. In both cases the possibility of intramolecular versus intermolecular reaction depends on the effective concentration of the reactants. The experimental methodology of model cyclic copolymers of styrene or perdeuterated styrene and butadiene is presented below [195]. This methodology is based on anionic polymerization-high vacuum techniques and controlled chlorosilane linking chemistry. The synthetic approach involved the reaction of a (1,3-phenylene)bis(3-methyl-1-phenylpentylidene)dilithium initiator with butadiene in the presence of sec-BuOLi, followed by the polymerization of styrene or perdeuterated styrene. The cyclization of the resulting α,ω-difunctional triblock copolymer was performed by using bis(dimethylchlorosilyl)ethane, under high dilution conditions. The copolymers were extensively characterized by SEC, NMR and UV spectroscopy, and membrane osmometry. The investigation of the micellar behavior of the cyclic copolymers in comparison with the triblock precursors by neutron scattering, revealed that the cyclic copolymers are very pure.

Anionic polymerization techniques were also critical for the synthesis of a model cyclic triblock terpolymer [cyclic(S-*b*-I-*b*-MMA)] [196]. The linear α,ω-amino acid precursor S-*b*-I-*b*-MMA was synthesized by the sequential anionic polymerization of St, I and MMA with 2,2,5,5-tetramethyl-1-(3-lithiopropyl)-1-aza-2,5-disilacyclopentane as the initiator and amine generator, and 4-bromo-1,1,1-trimethoxybutane as a terminator and carboxylic acid generator. Characterization studies of the intermediate materials as well as of the final cyclic terpolymer revealed high molecular and compositional homogeneity. Additional proof for the formation of the cyclic structure was provided by the lower intrinsic viscosity found for the cyclic terpolymer compared to that of the precursor.

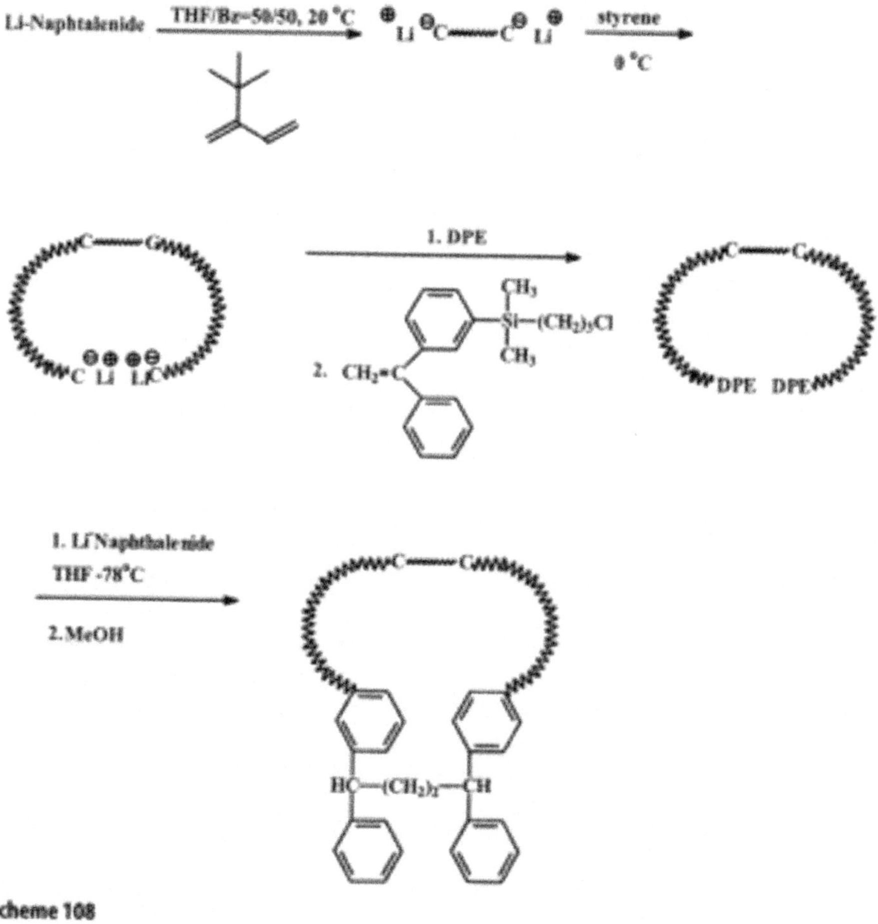

Scheme 108

In another case (Scheme 108) Li-naphthalenide served as the difunctional initiator for the subsequent polymerization first of 2-*tert*-butylbutadiene and subsequently of styrene [197]. The resulting difunctional Li(polystyrene-*b*-oligo(2-*tert*-butylbutadiene)-*b*-polystyrene)Li triblock precursor was reacted with DPE and was then terminated by 1-[3-(3-chloropropyldimethylsilyl) phenyl]-1-phenylethylene. Due to steric hindrance, the active sites of the difunctional living polymer react only with the chloropropyl groups of the terminator, resulting in a telechelic triblock copolymer with DPE-type vinyl groups at both ends. A cyclization reaction between the two vinyl groups was carried out by adding lithium naphthalenide to a dilute solution of the α,ω-DPE-functionalized triblock copolymer in a ratio of (DPE)/(Li) = 0.5, at − 78 °C. Lithium naphthalenide reacts with two DPE molecules to form the ring. The cyclic copolymer was isolated from the polycondensation impurities by preparative SEC. Ozonolysis of the diene moieties of the cyclic copolymer resulted in linear PS with almost the same molecular weight as the ring. On the other hand, ozonolysis of the linear precursor led to half the molecular weight of the ring, meaning that cyclic copolymer was successfully prepared.

4.5 Synthesis of Copolymers with Complex Macromolecular Architectures

Well-defined complicated macromolecular structures require complex synthetic procedures/techniques and characterization methods. Recently, several approaches leading to hyperbranched structures have been developed and will be the focus of this section. The preparation of hyperbranched poly(siloxysilane) has been reported [198] and is based on methylvinylbis(dimethyl siloxysilane), an A_2B type monomer, and a progressive hydrosilylation reaction with platinum catalysts. An appropriate hydrosilylation reaction on the peripheral – SiH groups led to the introduction of polymeric chain (PIB, PEO) or functional groups (epoxy, – NH_2) [199].

Hyperbranched polymers have also been prepared via living anionic polymerization. The reaction of poly(4-methylstyrene)-*b*-polystyrene lithium with a small amount of divinylbenzene, afforded a star-block copolymer with 4-methylstyrene units in the periphery [200]. The methyl groups were subsequently metalated with *s*-butyllithium/tetramethylethylenediamine. The produced anions initiated the polymerization of α-methylstyrene (Scheme 109). From the radius of gyration to hydrodynamic radius ratio (0.96–1.1) it was concluded that the second generation polymers behaved like soft spheres.

A combination of TEMPO living free radical (LFRP) and anionic polymerization was used for the synthesis of block-graft, block-brush, and graft-block-graft copolymers of styrene and isoprene [201]. The block-graft copolymers were synthesized by preparing a PS-*b*-poly(styrene-*co*-*p*-chloromethylstyrene) by LFRP [Scheme 110 (1)], and the subsequent re-

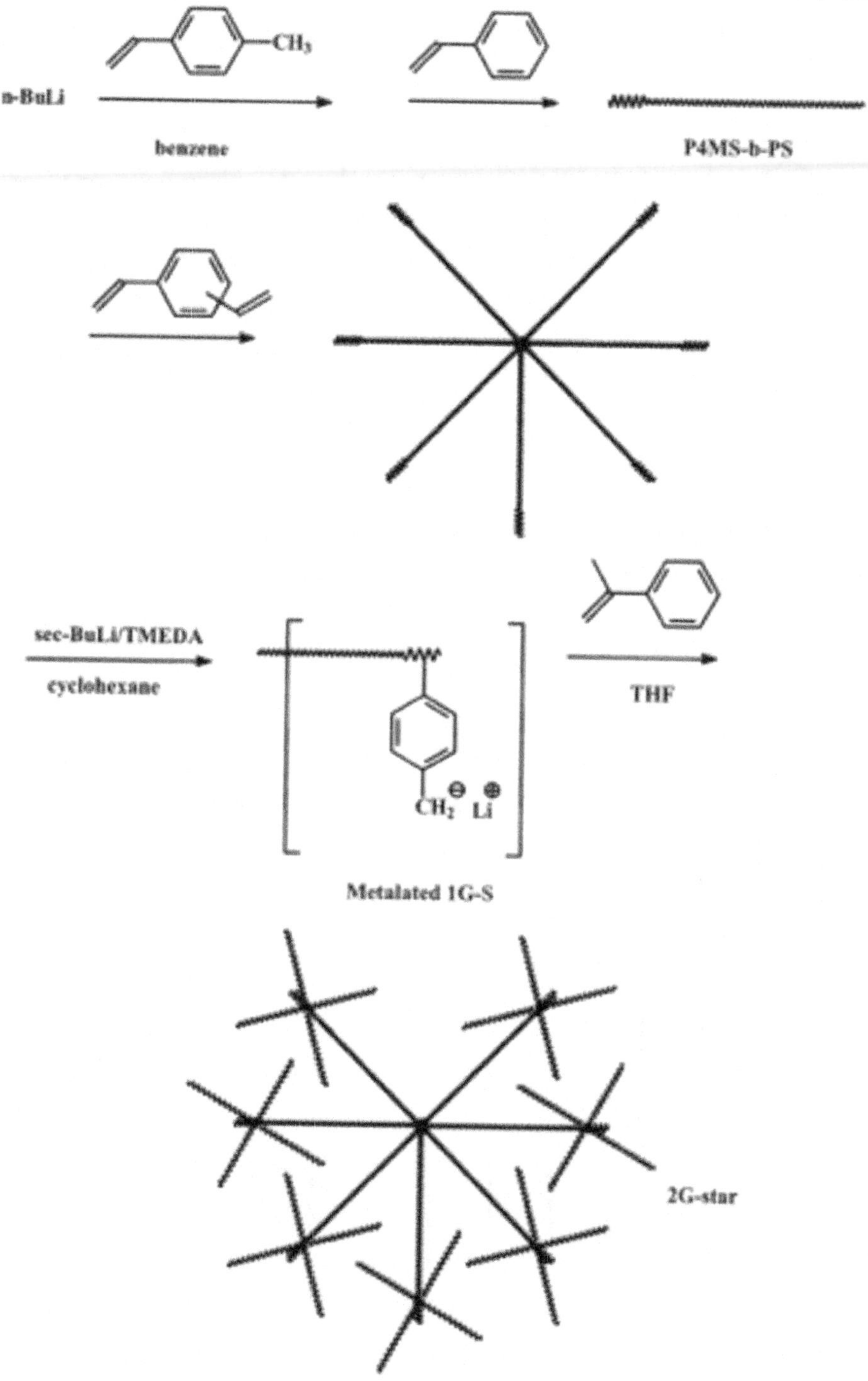

CH3
n-BuLi
benzene
P4MS-b-PS
sec-BuLi/TMEDA
cyclohexane
THF
CH2 Li
Metalated 1G-S
2G-star

Scheme 109

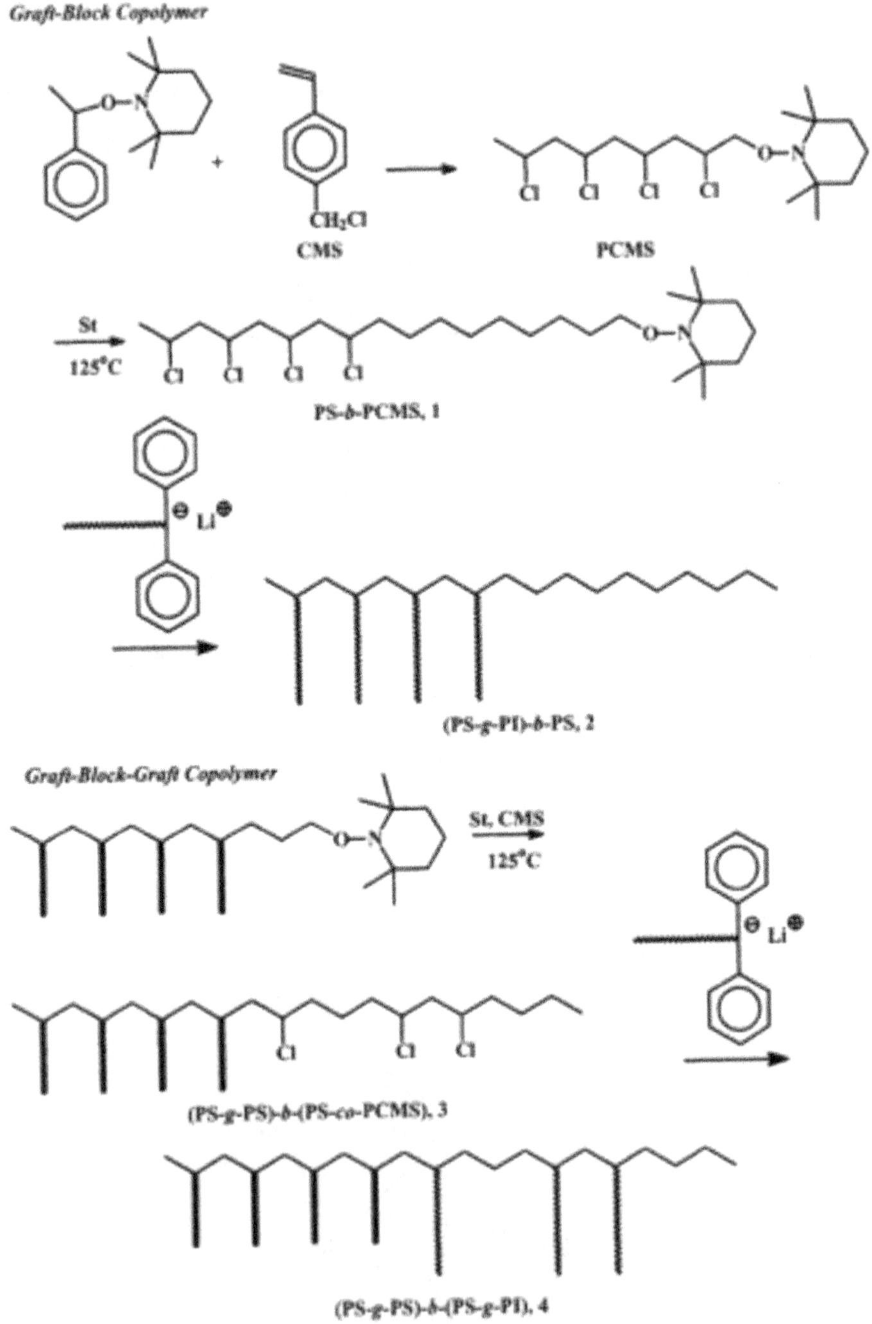
Graft-Block Copolymer
O
N
+
CH2Cl
CMS
Cl
PCMS
St
125°C
PS-b-PCMS, 1
Li
(PS-g-PI)-b-PS, 2
Graft-Block-Graft Copolymer
St, CMS
125°C
(PS-g-PS)-b-(PS-co-PCMS), 3
(PS-g-PS)-b-(PS-g-PI), 4

Scheme 110

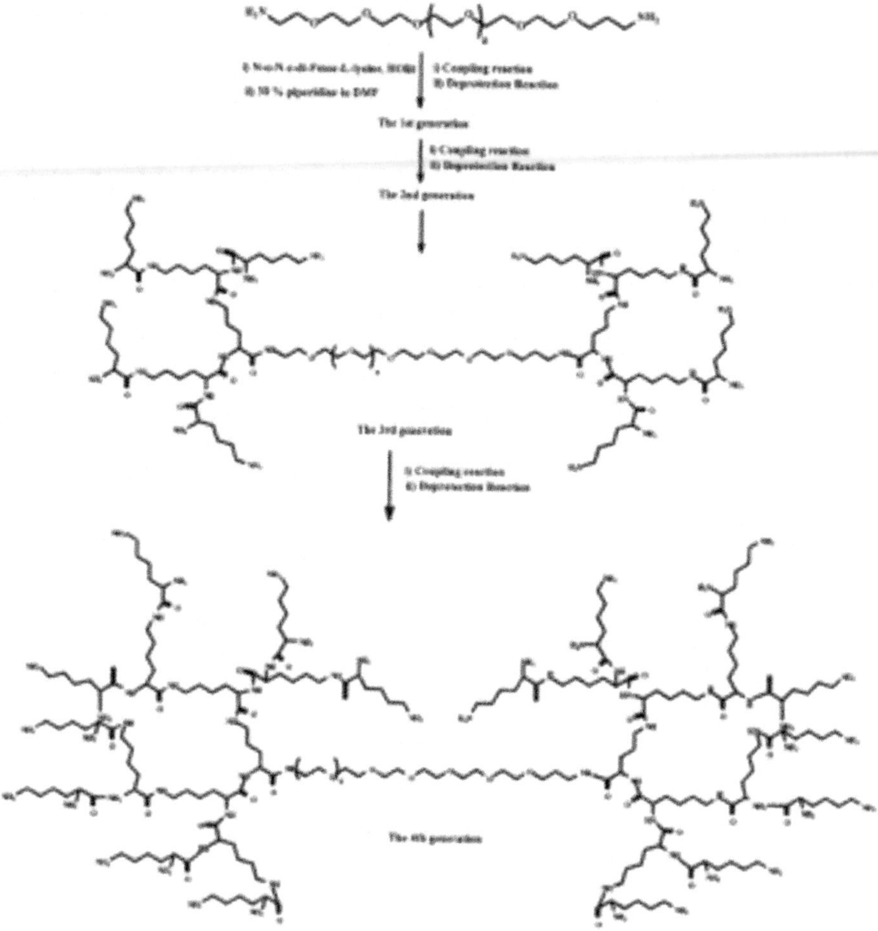

Scheme 111

action of the pendant chloromethyl groups with a low molecular weight 1,1-diphenylethylene end-capped polyisoprenyllithium (PI-DPELi) at – 20 °C. Under these conditions chlorine-lithium exchange reactions are minimized. The block-brush architecture was obtained in two steps. Initially, a PS-*b*-poly(*p*-chloromethylstyrene) (PS-*b*-PCMS) was synthesized by sequential LFRP of styrene and *p*-chloromethylstyrene. The *p*-chloromethyl groups of the diblock copolymers were reacted with the living PI chains, resulting in a block-brush copolymer [Scheme 110 (2)]. In addition, reaction of (PS-*g*-PS)-*b*-(PS-*co*-PCMS) [Scheme 110 (3)] with PI-DPELi resulted in a graft-block-graft copolymer [Scheme 110 (4)]. All products were analyzed and characterized by SEC, LALLS, membrane osmometry, NMR, and viscometry. In all cases molecular and compositional polydispersity was low (M_w/M_n = 1.08–1.32), and the linking efficiency was close to 100%.

A barbell-like ABA-type triblock copolymer, comprised of poly(L-lysine) (PLL) dendrimers (A) and poly(ethylene glycol) connector (B) has been reported [202]. The synthetic route involved the use of an α,ω-diamino-functionalized poly(ethylene glycol) as the polymeric supporter, for the attachment of $-NH_2$ protected lysine via an amidation reaction. The PLL dendrimer was generated at both ends of A by repeated liquid-phase peptide synthesis, as shown in Scheme 111. The intermediate products along with the final copolymers were characterized by MALDI-TOF MS. The results revealed that narrow molecular weight copolymers were synthesized having low molecular weights.

ATRP and "grafting from" methods led to the synthesis of poly(styrene-*g*-*tert*-butyl acrylate)-*b*-poly(ethylene-*co*-butylene)-*b*-poly(styrene-*g*-*tert*-butyl acrylate) block-graft copolymer [203]. ATRP initiating sites were produced along the PS blocks by chloromethylation as shown in Scheme 112. These sites then served to polymerize the *tert*-butyl acrylate. The poly(*tert*-butyl acrylate) grafts were hydrolyzed to result in the corresponding poly(acrylic

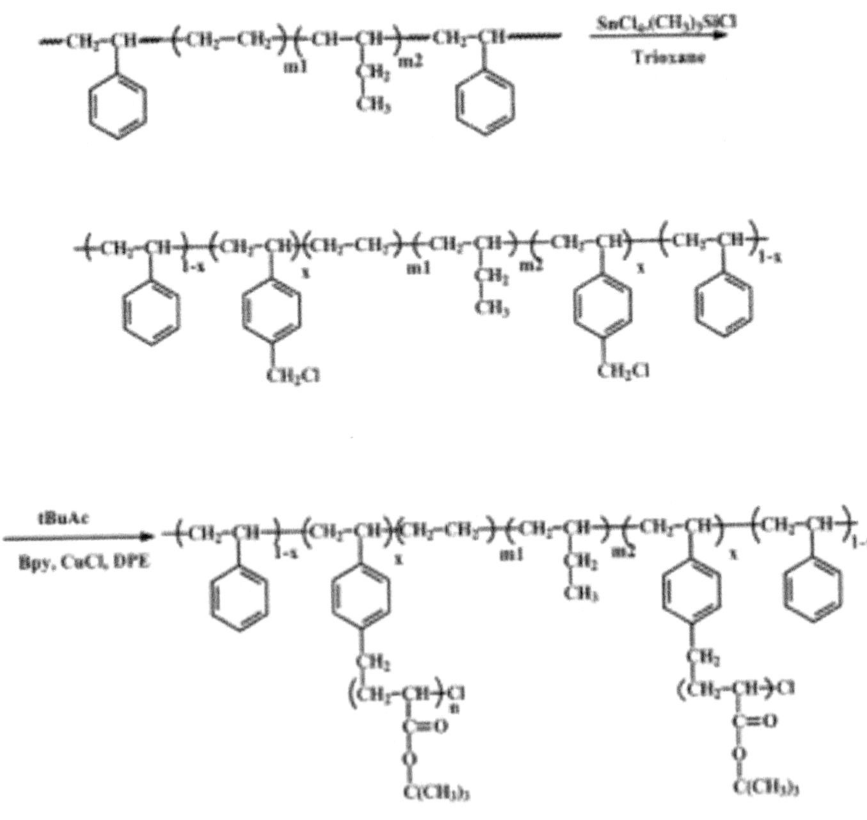

Scheme 112

acid) blocks, and their aggregation behavior in water was examined. The characterization of the block-grafts was conducted by using SEC and NMR.

The synthesis and characterization of PS-P2VP catenated block copolymers, i.e. two polymer rings held together solely by topological constraints, lacking a chemical or physical bond between the two rings has been reported [204]. The catenanes were prepared by end-to-end coupling of P2VP lithium dianion (synthesized by anionic polymerization with a difunctional initiator) with 1,4-bis(bromomethylenebenzene) (EX_2) in THF, in the presence of a PS macrocycle (Scheme 113). The isolation of the catenanes from the side products, PS and P2VP macrocycles, was performed by repeated extractions/centrifugations with methanol, which selectively dissolves the P2VP homopolymers. Characterization results from SEC revealed that the catenanes were successfully synthesized (Scheme 113).

Arborescent polystyrene homo- and PS-*g*-P2VP copolymers have been synthesized [205, 206]. The synthetic approach combines the repetitive anionic polymerization of styrene, the attachment of acetyl groups by partial acetylation of the benzene rings of styrene, and the reaction of living anionic polymeric chains with the acetyl groups. For the synthesis of the arborescent copolymers, in the final linking reaction P2VP Li chains were used. It was found that quantitative grafting occurs when the active sites were transformed to – 2VPLi instead of – SLi. The polymerizations along with the coupling reactions were conducted in a mixture of THF/toluene, at – 78 °C. This way, arborescent polymers with two generations and low polydispersity indices were synthesized.

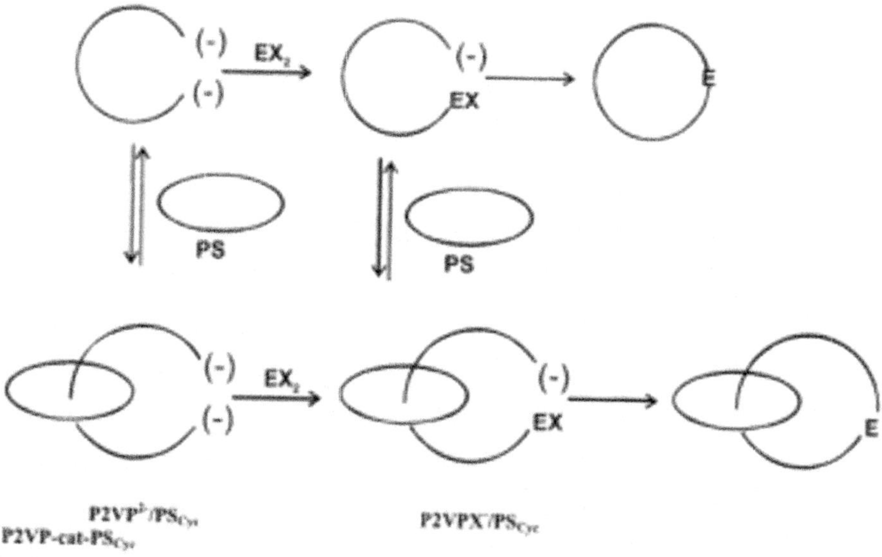

$P2VP^{2-}/PS_{Cyc}$ $P2VPX^{-}/PS_{Cyc}$
$P2VP\text{-}cat\text{-}PS_{Cyc}$

Scheme 113

Polystyrene/polyethylene oxide dendrimers were prepared by ATRP using tri- and tetra (bromomethyl) benzene as the initiators [207]. Each bromine end-group of the resulting stars was transformed first to two – OH groups and subsequently to potassium alcholate, as shown in Scheme 114. These – OK sites served to initiate the anionic polymerization of EO. The synthesized dendritic copolymers were found to display monomodal and narrow molecular weight distribution.

An iterative approach involving coupling reactions of living anionic polymers followed by functionalization, leads to three generation homo- and block copolymers. [208]. The reactions used are shown in Scheme 115.

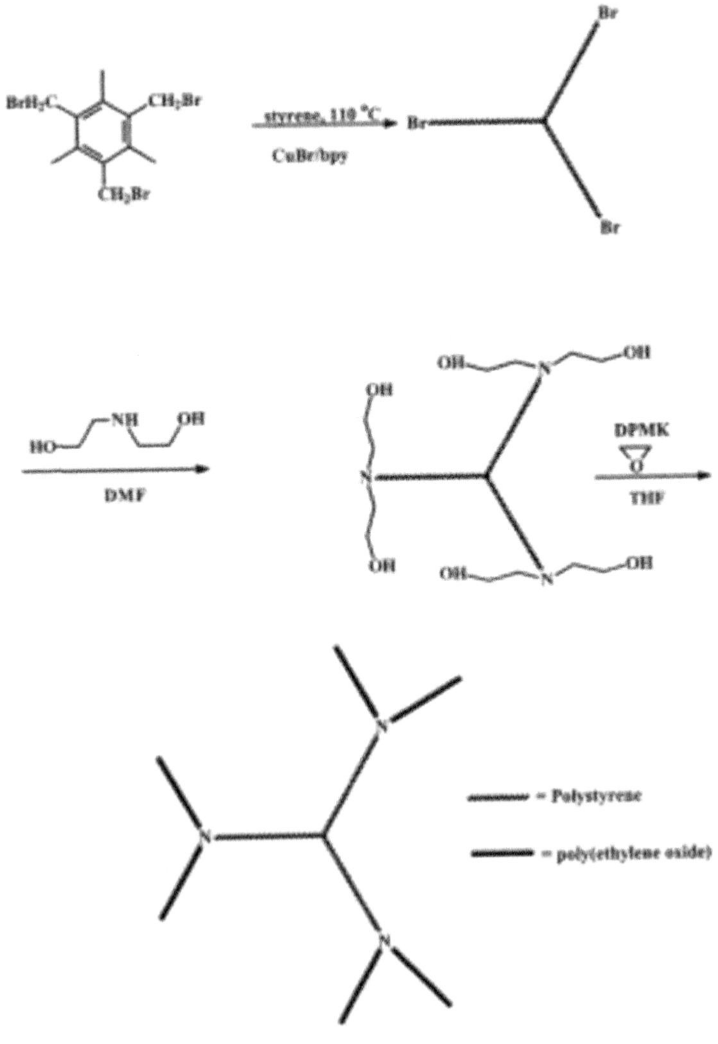

Scheme 114

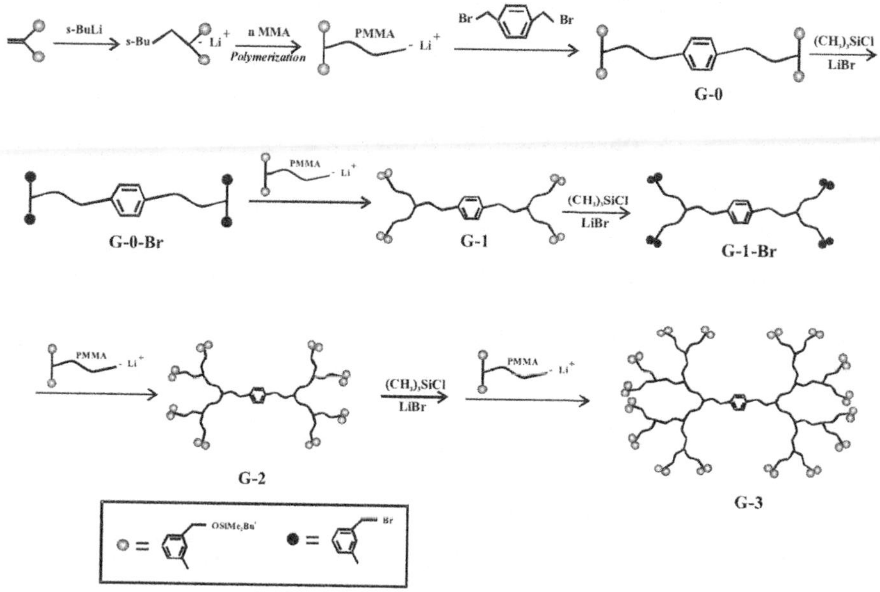

Scheme 115

The synthetic divergent approach involves the coupling reaction of α-functionalized bis(*tert*-butyldimethylsilyloxymethylphenyl PMMA lithium [(BDSIMP) PMMA] with dibromo xylene, resulting in an α,ω-tetra-functionalized PMMA with two BDSIMP groups at each end (G0). The BDSIMP groups are then transformed into benzyl bromide functionalities, followed by the reaction of another α-functionalized living anionic PMMA with two *tert*-butyldimethylsilyloxymethylphenyl (BDSIMP) groups (G1). This reaction sequence was successfully repeated twice to afford a series of homopolymers with up to three generations, all with well-defined architectures and precisely controlled chain lengths. Moreover, an amphiphilic dendrimer-like copolymer of PMMA and poly(2-hydroxyethyl methacrylate) (PHEMA) was synthesized in a similar manner, by using living PHEMA chains instead of living PMMA chains at the final reaction stage.

A final example is the synthesis of H-shaped copolymer of $(PS)_2PEG(PS)_2$ by ATRP, i.e. [209]. The synthetic strategy involves the synthesis of 2,2-bis(methylene α-bromopropionate) propionyl chloride (1), the preparation of 2,2-bis(methylene α-bromopropionate) propionyl-terminated poly(ethylene glycol) (BMBP-PEG-BMBP) (2), and then ATRP of styrene at 110 °C with BMBP-PEG-BMBP/CuBr/2,2′-bipyridine as the initiating system. The structure (3) was configured by using NMR and SEC measurements (Scheme 116).

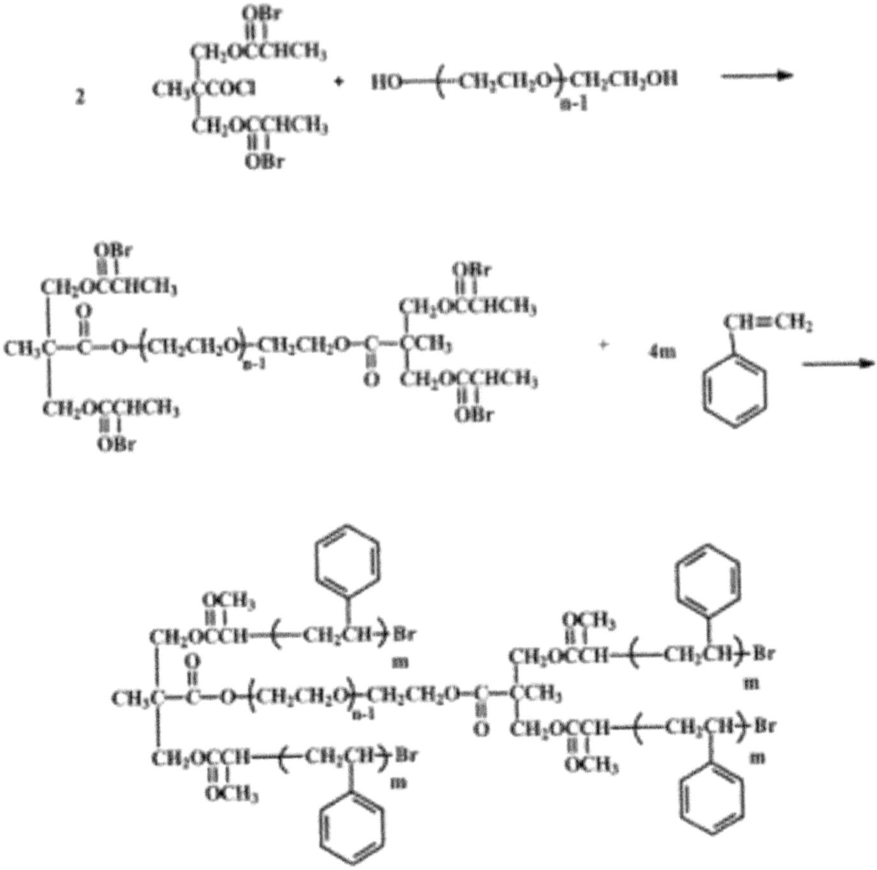

Scheme 116

5
Conclusions

Recent developments in polymer chemistry have allowed for the synthesis of a remarkable range of well-defined block copolymers with a high degree of molecular, compositional, and structural homogeneity. These developments are mainly due to the improvement of known polymerization techniques and their combination. Parallel advancements in characterization methods have been critical for the identification of optimum conditions for the synthesis of such materials. The availability of these well-defined block copolymers will facilitate studies in many fields of polymer physics and will provide the opportunity to better explore structure-property relationships which are of fundamental importance for hi-tech applications, such as high temperature separation membranes, drug delivery systems, photonics, multifunctional sensors, nanoreactors, nanopatterning, memory devices etc.

References

1. Hadjichristidis N, Pispas S, Floudas GA (2003) In: Block copolymers. Synthetic strategies, physical properties, and applications. Wiley, Weinheim
2. Pitsikalis M, Pispas S, Mays JW, Hadjichristidis N (1998) Adv Polym Sci 135:1
3. Lodge TP (2003) Macromol Chem Phys 204:265
4. Hsieh HL, Quirk RP (1996) In: Anionic polymerization. Principles and practical applications. Marcel Dekker, New York
5. Hong K, Urhig D, Mays JW (1999) Curr Opin Solid State Mater Sci 4:531
6. Hadjichristidis N, Pitsikalis M, Pispas S, Iatrou H (2001) Chem Rev 101:3747
7. Hadjichristidis N, Pispas S, Iatrou H, Pitsikalis M (2002) Curr Org Chem 6:155
8. Hadjichristidis N, Iatrou H, Pispas S, Pitsikalis M (2000) J Polym Sci Polym Chem Ed 38:3211
9. Tong JD, Jerôme R (2000) Polymer 41:2499
10. Tong JD, Leclére, Doneux C, Brédas JL, Lazzaroni R, Jerôme R (2000) Polymer 41:4617
11. Pitsikalis M, Siakali-Kioulafa E, Hadjichristidis N (2000) Macromolecules 33:5460
12. Pitsikalis M, Siakali-Kioulafa E, Hadjichristidis N (2004) J Polym Sci Polym Chem Ed 42:4177
13. Allen RD, Long TE, McGrath JE (1986) Polym Bull 15:127
14. Imae T, Tabuchi H, Funayama K, Sato A, Nakamura T, Amaya N (2000) Colloids Surf 167:73
15. Busse K, Kressler J, van Eck D, Höring S (2002) Macromolecules 35:178
16. de Paz Báñez MV, Robinson KL, Armes SP (2000) Macromolecules 33:451
17. Ishizone T, Tajima H, Torimae H, Nakahama S (2002) Macromol Chem Phys 203:2375
18. Lehmann O, Förster S, Springer J (2000) Macromol Rapid Commun 21:133
19. Zhang H, Ruckestein E (2001) Macromolecules 34:3587
20. Ishizone T, Han S, Okuyama S, Nakahama S (2003) Macromolecules 36:42
21. Quirk RP, Ma J-J (1988) J Polym Sci Polym Chem Ed 26:2031
22. Ekizoglou N, Hadjichristidis N (2001) J Polym Sci Polym Chem Ed 39:1198
23. Ekizoglou N, Hadjichristidis N (2002) J Polym Sci Polym Chem Ed 40:2166
24. Fragouli PG, Iatrou H, Hadjichristidis N (2002) Polymer 43:7141
25. Gohy J-F, Willet N, Varshney S, Zhang J-X, Jerôme R (2001) Angew Chem Int Ed 40:3214
26. Hückstädt H, Göpfert A, Abetz V (2000) Polymer 41:9089
27. Li H-J, Tsiang C-CR (2000) Polymer 41:5601
28. Lee J, Hogen-Esch TE (2001) Macromolecules 34:2805
29. Bellas V, Iatrou H, Hadjichristidis N (2000) Macromolecules 33:6993
30. Natori I (1997) Macromolecules 30:3696
31. Natori I, Inoue S (1998) Macromolecules 31:982
32. Natori I, Inoue S (1998) Macromolecules 31:4687
33. Imaizumi K, Ono T, Natori I, Sakurai S, Takeda K (2000) J Polym Sci Polym Phys Ed 39:13
34. Hong K, Mays JW (2001) Macromolecules 34:3540
35. Tsoukatos T, Avgeropoulos A, Hadjichristidis N, Hong K, Mays JW (2002) Macromolecules 35:7928
36. Williamson DT, Buchanan TD, Elkins CL, Long TE (2004) Macromolecules 37:4505
37. Muller M, Zentel R (1994) Macromolecules 27:4404
38. Shibaev PV, Tang K, Genack AZ, Kopp V, Green MM (2002) Macromolecules 35:3022
39. Maeda K, Okamoto Y (1998) Macromolecules 31:1046

40. Bur AJ, Fetters LJ (1976) Chem Rev 76:727
41. Lee JS, Ryu SW (1999) Macromolecules 32:2085
42. Shin YD, Kim SY, Ahn JH, Lee JS (2001) Macromolecules 34:2408
43. Ahn JH, Shin YD, Kim SY, Lee JS (2003) Polymer 44:3847
44. Ahn JH, Lee JS (2003) Macromol Chem Phys 24:571
45. Vazaios A, Pitsikalis M, Hadjichristidis M (2003) J Polym Sci Polym Chem Ed 41:3094
46. Schmalz H, Böker A, Lange R, Krausch G, Abetz V (2001) Macromolecules 34:8720
47. Schmalz H, Knoll A, Müller, Abetz V (2002) Macromolecules 35:10004
48. Matyjaszewski K (ed) (1996) Cationic polymerization. Mechanisms, synthesis and applications. Marcel Dekker, New York
49. Kennedy JP, Iván B (1992) Designed polymers by carbocationic macromolecular engineering. Theory and practice. Hanser Publishers, Munchen, Germany
50. Faust R, Shaffer TD (eds) (1997) Cationic polymerization. Fundamentals and applications. ACS Symp Series 666
51. Ebdon JR, Eastmond GC (eds) (1995) New methods of polymer synthesis, Chap 2. Blackie Academic and Professional, Glasgow, UK
52. Hatada K, Kitayama T, Vogl O (eds) (1997) Macromolecular design of polymeric materials, Chap 3 and 4. Marcel Dekker, New York
53. Verdonck B, Goethals EJ, Du Prez FE (2003) Macromol Chem Phys 204:2090
54. Matsumoto K, Mazaki H, Matsuoka H (2004) Macromolecules 37:2256
55. Satoh K, Kamigaito M, Sawamoto M (2000) Macromolecules 33:5830
56. Shibasaki Y, Sanada H, Yokoi M, Sanda F, Endo T (2000) Macromolecules 33:4316
57. Allock HR, Reeves SD, Nelson JM, Manners I (2000) Macromolecules 33:3999
58. Wang Y, Goethals EJ (2000) Macromolecules 33:808
59. Odian G (1991) Principles of polymerization, 3rd ed. Wiley, New York
60. Hawker CJ, Bosman AW, Harth E (2001) Chem Rev 101:3661
61. Matyjaszewski K (ed) (1998) Controlled radical polymerization. ACS Symposium Series, Chap 10–15
62. Matyjaszewski K, Xia J (2001) Chem Rev 101:2921
63. Coessens V, Pintauer T, Matyjaszewski K (2001) Progr Polym Sci 26:337
64. Matyjaszewski K (ed) (1998) Controlled radical polymerization. ACS Symposium Series, Chap 16–21
65. Goto A, Fukuda T (2004) Progr Polym Sci 29:329
66. McCormick CJ, Lowe AB (2004) Acc Chem Res 37:312
67. Yousi Z, Jian L, Ronhchuan Z, Jianliang Y, Lizong D, Lansum Z (2000) Macromolecules 33:4745
68. Benoit D, Harth E, Fox P, Waymouth RM, Hawker CJ (2000) Macromolecules 33:363
69. Chalari I, Pispas S, Hadjichristidis N (2001) J Polym Sci Polym Chem Ed 39:2889
70. Gopalan P, Zhang Y, Li X, Wiesner U, Ober CK (2003) Macromolecules 36:3357
71. Diaz T, Fischer A, Jonquières A, Brembilla A, Lochon P (2003) Macromolecules 36:2235
72. Gan L-H, Ravi P, Mao BW, Tam K-C (2003) J Polym Sci Polym Chem Ed 41:2688
73. Narrainen PA, Pascual S, Haddleton DM (2002) J Polym Sci Polym Chem Ed 40:439
74. Yang R, Wang Y, Wang X, He W, Pan C (2003) Eur Polym J 39:2029
75. Davis KA, Matyjaszewski K (2001) Macromolecules 34:2101
76. Moineau G, Minet M, Teyssié P, Jérôme R (2000) Macromol Chem Phys 201:1108
77. Masci G, Bontempo D, Tiso N, Diociaiuti M, Mannina L, Capitani D, Crescenzi V (2004) Macromolecules 37:4464
78. Lu S, Fan Q-L, Liu S-Y, Chua S-J (2002) Macromolecules 35:9875

79. Onishi I, Baek K-Y, Kotani Y, Kamigaito M, Sawamoto M (2002) J Polym Sci Polym Chem Ed 40:2033
80. Sumerlin BS, Donovan MS, Mitsukami Y, Lowe AB, McCormick CL (2001) Macromolecules 34:6561
81. Taton A, Wilczewska A-Z, Destarac M (2001) Macromol Rapid Commun 22:1497
82. Arotçaréna M, Heise B, Ishaya S, Laschewsky A (2002) J Am Chem Soc 124:3787
83. Sumerlin BS, Lowe AB, Thomas DB, Convertine AJ, Donovan MS, McCormick CL (2004) J Polym Sci Polym Chem Ed 42:1724
84. Webster OW, Hertler WR, Sogah DY, Farnham WB, Rajanbabu TV (1983) J Am Chem Soc 105:5706
85. Webster OW, Anderson BC, Mijs WJ (ed) (1992) New methods for polymer synthesis. Plenum Press, New York
86. Hatada K, Kitayama T, Vogl O (eds) (1997) Macromolecular design of polymeric materials, Chap 7. Marcel Dekker, New York
87. Bütün V, Armes SP, Billingham NC (2001) Polymer 42:5993
88. Bütün V, Vamvakaki M, Billingham NC, Armes SP (2000) Polymer 41:3173
89. Triftaridou AI, Vamvakaki M, Patrickios CS (2002) Polymer 43:2921
90. Buchmeiser MR (2000) Chem Rev 100:1565
91. Ofstead EA, Wagener KB, Mijs WJ (ed) (1992) New methods for polymer synthesis. Plenum Press, New York
92. Trnka TM, Grubbs RH (2001) Acc Chem Res 34:18
93. Grubbs RH, Novak BM (1990) Encyclopedia of polymer science and engineering, Suppl I. Wiley, New York, p 20
94. Grubbs RH, Tumas W (1989) Science 243:907
95. Wagener KB, Boncella JM, Duttweiler RP, Nel JG (1989) Polym Prepr Am Chem Soc 30(1):283
96. Wagener KB, Boncella JM, Nel JG (1991) Macromolecules 24:2649
97. Ebdon JR, Eastmond GC (eds) (1995) New methods of polymer synthesis, Chap 3. Blackie Academic and Professional, Glasgow, UK
98. Trzaska ST, Lee L-BW, Register RA (2000) Macromolecules 33:9215
99. Bazzi HS, Sleiman HF (2002) Macromolecules 35:9617
100. Liaw D-J, Tsai C-H (2000) Polymer 41:2773
101. Chen B, Sleiman HF (2004) Macromolecules 37:5866
102. Miyamoto Y, Fujiki M, Nomura K (2004) J Polym Sci Polym Chem Ed 42:4248
103. Lehn J-M (1995) Supramolecular chemistry, concepts and perspectives. Wiley, Weinheim, Germany
104. Schubert US, Eschbaumer C (2002) Angew Chem Int Ed 41:2892
105. Lohmeijer BGG, Schubert US (2003) J Polym Sci Polym Chem Ed 41:1413
106. Andres PR, Schubert US (2004) Adv Mater 16:1043
107. Gohy J-F, Lohmeijer BGG, Schubert US (2002) Macromolecules 35:4560
108. Gohy J-F, Lohmeijer BGG, Schubert US (2002) Macromol Rapid Commun 23:555
109. Gohy J-F, Lohmeijer BGG, Varshney SK, Décamps B, Leroy E, Boileau, Schubert US (2002) Macromolecules 35:9748
110. Hustad PD, Coates GW (2002) J Am Chem Soc 124:11578
111. Barbier-Baudry D, Bonnet F, Dormond A, Finot E, Visseaux M (2002) Macromol Chem Phys 203:1194
112. Frauenrath H, Balk S, Keul H, Höcker H (2001) Macromol Rapid Commun 22:1147
113. Batis C, Karanikolopoulos G, Pitsikalis M, Hadjichristidis N (2003) Macromolecules 36:9763
114. Cameron PA, Gibson VC, Graham AJ (2000) Macromolecules 33:4329

115. Kakehi T, Yamashita M, Yasuda H (2000) React Funct Polym 46:81
116. Akiyama Y, Harada A, Nagasaki Y, Kataoka K (2000) Macromolecules 33:5841
117. Wang Y, Hillmyer MA (2000) Macromolecules 33:7395
118. Castle TC, Hutchings LR, Khosravi E (2004) Macromolecules 37:2035
119. Xu Y, Pan C (2000) J Polym Sci Polym Chem Ed 38:337
120. Tao L, Luan B, Pan C-Y (2003) Polymer 44:1013
121. Luo L, Ranger M, Lessard DG, Le Garrec D, Gori S, Leroux J-C, Rimmer S, Smith D (2004) Macromolecules 37:4008
122. Bielawski CW, Morita T, Grubbs RH (2000) Macromolecules 33:678
123. Matsugi T, Kojoh S-I, Kawahara N, Matsuo S, Kaneko H, Kashiwa N (2003) J Polym Sci Polym Chem Ed 41:3965
124. Ranger M, Jones M-C, Yessine M-A, Leroux J-C (2001) J Polym Sci Polym Chem Ed 39:3861
125. Reining B, Keul H, Höcker H (2002) Polymer 43:7145
126. Narain R, Armes SP (2003) Biomacromolecules 4:1746
127. Liu S, Billingham NC, Armes SP (2001) Angew Chem Int Ed 40:2328
128. Liu S, Armes SP (2001) J Am Chem Soc 123:9910
129. Robinson KL, de Paz-Báñez MV (2001) Macromolecules 34:5799
130. Resendes R, Massey J, Dorn H, Winnik MA, Manners I (2000) Macromolecules 33:8
131. Loos K, Müller AHE (2002) Biomacromolecules 3:368
132. Kurian P, Zschoche S, Kennedy JP (2000) J Polym Sci Polym Chem Ed 38:3200
133. Hua FJ, Yang YL (2001) Polymer 42:1361
134. Bernaerts KV, Schacht EH, Goethals EJ, du Prez FE (2003) J Polym Sci Polym Chem Ed 41:3206
135. Tunca U, Karliga B, Ertekin S, Ugur AL, Sirkecioglu O, Hizal G (2001) Polymer Commun 42:8489
136. Tunca U, Erdogan T, Hizal G (2002) J Polym Sci Polym Chem Ed 40:2025
137. Smith AP, Fraser CL (2003) Macromolecules 36:2654
138. Meyer U, Palmans ARA, Lootjens T, Heise A (2002) Macromolecules 35:2873
139. Wang C, Cui M (2003) J Appl Polym Sci 88:1632
140. Jeon O, Lee S-H, Kim SH, Lee YM, Kim YH (2003) Macromolecules 36:5585
141. Ahn C-H, Chae SY, Bae YH, Kim SW (2004) J Control Release 97:567
142. Hadjichristidis N, Pispas S, Pitsikalis M, Iatrou H, Vlahos C (1999) Adv Polym Sci 142:71
143. An SG, Cho CG (2004) Macromol Rapid Commun 25:618
144. Peetz RM, Moustafa AF, Kennedy JP (2003) J Polym Sci Polym Chem Ed 41:740
145. Kwon Y, Puskas JE (2004) Eur Polym J 40:119
146. Yijin X, Caiyuan P (2000) Macromolecules 33:4750
147. Narita M, Nomura R, Tomita I, Endo T (2000) Macromolecules 33:4979
148. Angot S, Taton D, Gnanou Y (2000) Macromolecules 33:5418
149. Francis R, Taton D, Logan JL, Masse P, Gnanou Y, Duran RS (2003) 36:8253
150. Hou S, Chaikof EL, Taton D, Gnanou Y (2003) Macromolecules 36:3874
151. Lele BS, Leroux J-C (2002) Polymer 43:5595
152. Park SY, Han BR, Na KM, Han DK, Kim SC (2003) Macromolecules 36:4115
153. Hong K, Wan Y, Mays JW (2001) Macromolecules 34:2482
154. Mignard E, Hiorns RC, François B (2002) Macromolecules 35:6132
155. Bosman AW, Vestberg R, Heumann A, Fréchet JMJ, Hawker CJ (2003) J Am Chem Soc 125:715
156. Yang JC, Mays JW (2002) Macromolecules 35:3433
157. Baek K-Y, Kamigaito M, Sawamoto M (2002) J Polym Sci Polym Chem Ed 40:633

158. Héroguez V, Amédro E, Grande D, Fontanille M, Gnanou Y (2000) Macromolecules 33:7241
159. Johnson RM, Fraser CL (2004) Macromolecules 37:2718
160. He T, Li D, Sheng X, Zhao B (2004) Macromolecules 37:3128
161. Celik C, Hizal G, Tunca U (2003) J Polym Sci Part A Pol Chem 41:2542
162. Guo Y, Xu J, Pan C (2001) J Polym Sci Part A Pol Chem 39:437
163. Hadjichristidis N (1999) J Polym Sci Part A Pol Chem 37:857
164. Tsoukatos T, Hadjichristidis N (2002) J Polym Sci Part A Pol Chem 40:2575
165. Mavroudis A, Avgeropoulos A, Hadjichristidis N, Thomas E, Lohse D (2003) Chem Mater 15:1976
166. Zhu Y, Gido S, Moschakou M, Iatrou H, Hadjichristidis N, Park S, Chang T (2003) Macromolecules 36:5719
167. Eschwey H, Burchard W (1975) Polymer 16:180
168. Tsitsilianis C, Chaumont P, Rempp P (1990) Makromol Chem 191:2319
169. Tsitsilianis C, Graff S, Rempp P (1990) Eur Polym J 27:243
170. Tsitsilianis C, Lutz P, Graff S, Lamps J-P, Rempp P (1991) Macromolecules 24:5897
171. Rein D, Rempp P, Lutz P (1993) J Makromol Chem Macromol Symp 67:237
172. Höcker H, Latterman G (1976) J Polym Sci Symp 54:361
173. Quirk P, Yoo T, Lee B (1994) JMS Pure Appl Chem A31(8):911
174. Yoo T (1994) PhD Thesis, Akron University
175. Bae C, Fodor Z, Faust R (1997) Macromolecules 30:198
176. Bae C, Faust R (1997) Polym Prepr ACS Div Polym Chem 39:621
177. Fernyhough C, Young R (2000) Macromol Symp 161:103
178. Hirao A, Tokuda Y, Morifuji K, Hayashi M (2001) Macromol Chem Phys 202:1606
179. Lambert O, Reutenauer S, Hurtez G, Dumas P (2000) Macromol Symp 161:97
180. Dumas P, Delaite C, Hurtrez G (2002) Macromol Symp 183:29
181. Fraser C, Smith A, Wu X (2000) J Am Chem Soc 122:9026
182. Shi P, Li Y, Pan C (2004) Eu Polym J 40:1283
183. Li Y, Wang Y, Pan C (2003) J Polym Sci Part A Pol Chem 41:1243
184. Schubert U, Hohmeier H (2002) Macromol Rapid Commun 23:561
185. Liu S, Sen A (2001) Macromolecules 34:1529
186. Lu J, Liang H, Li A, Cheng Q (2004) Eu Polym J 40:397
187. Chang J, Park P, Han M (2000) Macromolecules 33:321
188. Ming G, Ting W, Yin-Fang Z, Cai-Yan P (2001) Polymer 42:6385
189. Schulze U, Fonagy T, Komber H, Pompe G, Pionteck J, Ivan B (2003) Macromolecules 36:4719
190. Endo K, Sugita T (2004) J Polym Sci Part A Pol Chem 42:2904
191. Zhang D, Ortiz C (2004) Macromolecules 37:4271
192. Pantazis D, Chalari I, Hadjichristidis N (2003) Macromolecules 36:3783
193. Uhrig D, Mays J (2002) Macromolecules 35:7182
194. Iatrou H, Mays J, Hadjichristidis N (1998) Macromolecules 31:6697
195. Iatrou H, Hadjichristidis N, Meier G, Frielinghoaus H, Monkenbush M (2002) Macromolecules 35:5426
196. Pantazis D, Schultz D, Hadjichristidis N (2002) J Polym Sci Part A Pol Chem 40:1476
197. Takano A, Nonaka A, Kadoi O, Hirahara K, Kawahara S, Isono Y, Torikai N, Matsushita Y (2002) J Polym Sci Part B Pol Phys 40:1582
198. Gong G, Miravet J, Frechet J (1999) J Polym Sci Part A Pol Chem 37:3193
199. Gong G, Miravet J, Frechet J (2000) J Polym Sci Part A Pol Chem 38:2970
200. Ishizu K, Takahashi D, Takeda H (2000) 41:6081
201. Tsoukatos T, Pispas S, Hadjichristidis N (2000) Macromolecules 33:9504

202. Choi J, Joo D, Kim C, Kim K, Park J (2000) J Am Chem Soc 122:474
203. Ning F, Jiang M, Mu M, Duan H, Xie J (2002) J Polym Sci Part A Pol Chem 40:1253
204. Gan Y, Dong D, Hogen-Esch T (2002) Macromolecules 35:6799
205. Gauthier M, Li J, Dockendorf J (2003) Macromolecules 36:2462
206. Li J, Gauthier M (2001) Macromolecules 34:8918
207. Francis R, Taton D, Logan J, Masse P, Gnanou Y, Duran R (2003) Macromolecules 36:8253
208. Matsuo A, Watanabe T, Hirao A (2004) Macromolecules 37:6283
209. Li Y, Shi P, Pan C (2004) Macromolecules 37:5190

Adv Polym Sci (2005) 189: 125–212
DOI 10.1007/12_004

Published online: 25 October 2005

Phase Behaviour and Morphologies of Block Copolymers

Volker Abetz (✉) · Peter F. W. Simon

Institut für Polymerforschung, GKSS-Forschungszentrum Geesthacht GmbH,
Max-Planck-Strasse 1, 21502 Geesthacht, Germany
volker.abetz@gkss.de, peter.simon@gkss.de

Abstract This chapter gives an overview of the research on the self-assembly of amorphous block copolymers at different levels of hierarchy. Besides the influence of composition and topology on the morphologies of block copolymers with linear, cyclic and branched topologies blends of block copolymers with low molecular weight components, other polymers or block copolymers and nanoparticles will also be presented.

Keywords Self-Assembly · Blend · Block copolymer · Microphase separation · Star copolymer · Superlattice

Abbreviations

a	scaling exponent (arbitrary)
A15	sphere phase, $Pm\bar{3}n$ symmetry
A	prefactor (arbitrary)
as	asymetric
-b-	*-block-*
bcc	body-centred cubic
C	hexagonal array of cylinders; symmetry: $p6mm$; alternatively abbreviated *Hex*
C14	*n*-tetradecane
CL	crystalline lamellar
-co-	*-copolymer-*
cmt	critical micelle temperature
CSC	core-shell cylinders
CSG	core-shell gyroid
D	diffussion coefficient
D	lamellar spacing
DBP	di-*n*-butyl phthalate
DEP	diethyl phthalate
Dis	disordered state
DMT	demicellization temperature
DMTA	dynamic mechanical thermal analysis
DOP	dioctyl phthalate
e	as subscript: entanglement
f	volume fraction
fcc	face-centred cubic spheres
G	double gyroiod phase; symmetry: $Ia\bar{3}d$; alternatively abbreviated: *Gyr*
G_i	dendron of generation *i*
G'	storage modulus
G''	loss modulus
Gyr	double gyroiod phase; symmetry: $Ia\bar{3}d$; alternatively abbreviated *G*

hcp	hexagonally closed packing
HPLC	high-performance liquid chromatography
hex	hexagonal array of cylinders; symmetry: *p6mm*; alternatively abbreviated *C*
L	lamellar phase; alternatively abbreviated *lam*
L	length in general
lam	lamellar phase alternatively abbreviated *L*
LAM_2	two-domain lamellae structure
LAM_3	three-domain lamellae structure
LCST	lower critical solution temperature
LDOT	lower disorder-to-order transition
LDT	lattice disordering transition
M_n	number average molecular weight
M_w	weight average molecular weight
MFT	mean-field theory
MDA	4,4′-methylenedianiline
N	degree of polymerization
ODT	order-disorder transition
OOT	order-order transition
P2MP	poly(2-methyl-1,3-pentadiene)
P2VP	poly(2-vinylpridine)
PA	polyamide
PB	polybutadiene
PBut	poly(1,2-butylene)
PCEMA	poly(2-cinnamoyloxyethyl methacrylate)
PCHMA	poly(cyclohexyl methacrylate)
PDMS	polydimethylsiloxane
PEB	poly(ethylene-*alt*-butylene)
PEO	poly(ethylene oxide)
PEE	poly(ethyl ethylene)
PEP	poly(ethylene-*alt*-propylene)
PI	polyisoprene
PL	perforated lamellar phase
PLA	poly(*D,L*-lactide)
PLLA	poly(*L*-lactide)
PLS	pillared lamellar structure
PMMA	poly(methyl methacrylate)
P*n*PMA	poly(*n*-pentyl methacrylate)
PPE	poly(2,6-dimethylphenylene ether)
PS	polystyrene
P*t*BA	poly(*tert*-butyl acrylate)
P*t*BMA	poly(*tert*-butyl methacrylate)
PTMSS	poly(4-trimethylsilylstyrene)
q	position of reflection peak
r	ratio of asymmetric to symmetric PS-*b*-PI diblock
-ran-	*-random-*
R_g	radius of gyration
s	symmetric
S	sphere phase (bcc array)
SAN	poly(styrene-*co*-acrylonitrile)
SAXS	small-angle X-ray scattering

SBS	polystyrene-*b*-polybutadiene-*b*-polystyrene
SCFT	self-consistent field theory
SEC	size exclusion chromatograpy
SPL	semiperforated lamellae
SSL	strong segregation limit
SST	strong segregation theory
-stat-	*-statistical-*
T	temperature (general)
T_{DMT}	demicellization temperature
T_g	glass transition temperature
T_{ODT}	order-disorder transition temperature
T_{OOT}	order-order transition temperature
TEM	transmission electron microscopy
UODT	upper order-to-disorder transition
v	crystall growth rate
WSL	weak segregation limit
ϕ	volume fraction of the polymer
χ	Flory-Huggins segment-segment interaction parameter
ρ	spacing ratio
τ	asymmetry parameter

1 Introduction

Block copolymers are composed of different polymer chains. Depending on the number of different blocks, their composition, and the way they are linked together (linear, branched, star-like or cyclic), they can form a variety of ordered structures with characteristic lengths on the mesoscale. This self-organization into ordered structures (also called morphologies) has attracted considerable interest among theoreticians and experimentalists during the last few decades, as indicated by an increasing number of publications in this field. Living anionic polymerization enabeled the controlled synthesis of block copolymers, and the number of accessible monomers since its inception has increased due to further developments in various types of controlled polymerization techniques. This chapter will deal with the morphological properties of block copolymers. There have been many excellent reviews on this topic [1–7], and for that reason the focus of this contribution will be on the developments in this area during the last few years. Due to the large number of interesting publications in the field, we cannot include all of them. Our goal is to cover the different aspects in the research on the phase behaviour and morphology of block copolymers, basically in the bulk state. These works are summarized in a section on linear-block copolymers and in a section on block copolymers with other topologies. Besides the pure block copolymers, their blends with other components are also fascinating research areas. Starting with changing thermodynamics by temperature variation or by adding low molecular weight solvents to block copolymers, homopolymers or other block copolymers, a large field has opened up for morphological transitions, i.e. manipulating block copolymer morphologies.

2
Linear Block Copolymers

The best-known and simplest class of block copolymers are linear diblock copolymers (AB). Being composed of two immiscible blocks, A and B, they can adopt the following equilibrium microphase morphologies, basically as a function of composition: spheres (*S*), cylinders (*C* or *Hex*), double gyroid (*G* or *Gyr*), lamellae (*L* or *Lam*), cf. Fig. 1 and the inverse structures. With the exception of the double gyroid, all morphologies are ideally characterized by a constant mean curvature of the interface between the different microdomains.

For nearly symmetric compositions the unlike blocks form domains composed of alternating layers, known as lamellar phase (*L*). Slightly off-symmetry composition results in the formation of a different layered structure. The structure is known as perforated layers (*PL*) or catenoid phase. Despite an earlier assignment as an equilibrium phase, it is now known to be in a long-lived metastable state that facilitates the transition from *L* to *G* phases [9–14]. The *PL* structure consists of alternating minority and majority component layers in which hexagonally packed channels of the majority component extend through the minority component.

A sketch of the perforated lamellar morphology is depicted in Fig. 2. Depending on the point of view, the 2D projection exhibits different patterns: perpendicular to the layers of the perforated lamellar structure the projection appears like a hexagonal honeycomb mesh (Fig. 2a). By contrast, the parallel view (Fig. 2b) leads to rows of spots.

The *G* phase is a complex phase possessing a bicontinuous structure which is located in or close to the compositional range of the *PL* phase. The minority component forms two interpenetrating threefold coordinated networks [10].

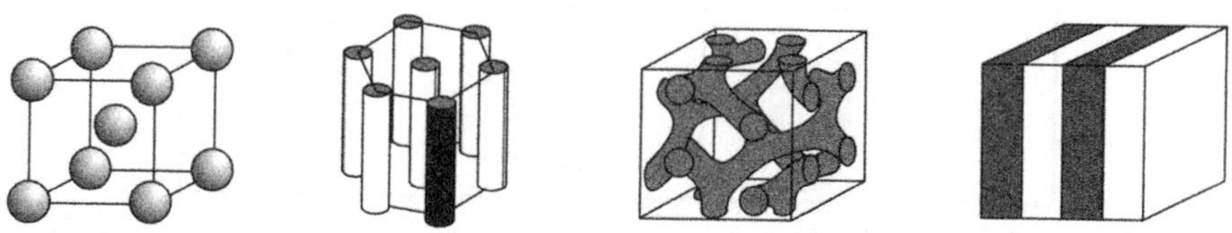

Fig. 1 Morphologies of diblock copolymers: cubic packed spheres (*S*), hexagonal packed cylinders (*C* or *Hex*), double gyroid (*G* or *Gyr*), and lamellae (*L* or *Lam*). Inverse phases not shown. From [8]. Copyright 2000 Wiley

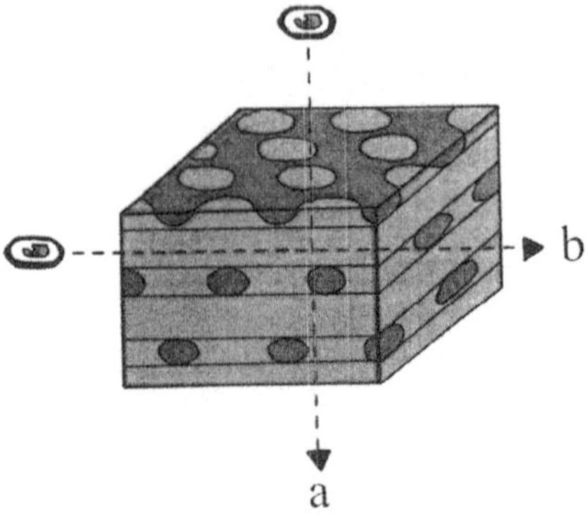

Fig. 2 Sketch of perforated layers (*PL*) or catenoid structure. **a** Projection direction perpendicular to layers of perforated lamellar structure appears as a hexagonal honeycomb mesh. **b** Projection parallel to layers appears as *rows of dark spots* resulting from cross section of parts. Sketch according to [15]. Copyright 2001 American Chemical Society

When the compositional asymmetry is further increased, the minority component assembles in hexagonally packed cylinders (*C*). Finally, it is organized in an array of spheres, cf. Table 1. A body-centred cubic lattice arrangement (*S* or *bcc*) is mostly observed; however, other symmetries like the face-centred (*fcc*) or A15 cubic were also reported and will be reviewed in Sects. 8.3 and 7.4 respectively.

Even more complex structures have been described. For example, chirality of blocks may lead to other morphologies. A polystyrene-*b*-poly-(*L*-lactide) diblock copolymer, PS-*b*-PLLA, constituting both achiral and chiral blocks was reported to form an array of hexagonally packed PLLA nanohelices with a left-handed helical sense in the bulk state (Fig. 3). The structure was found

Table 1 Different arrays of spheres

Symmetry	*Fcc*	*Bcc*	A15
Space group	$Fm\bar{3}n$	$Im\bar{3}m$	$Pm\bar{3}n$
Sketch			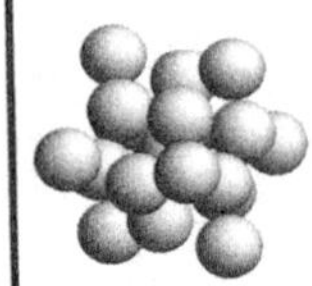

Pictures taken from Crystal Lattice Structures Web page: cst-www.nrl.navy.mil/lattice/ provided by Center for Computational Materials Science of United States Naval Research Laboratory

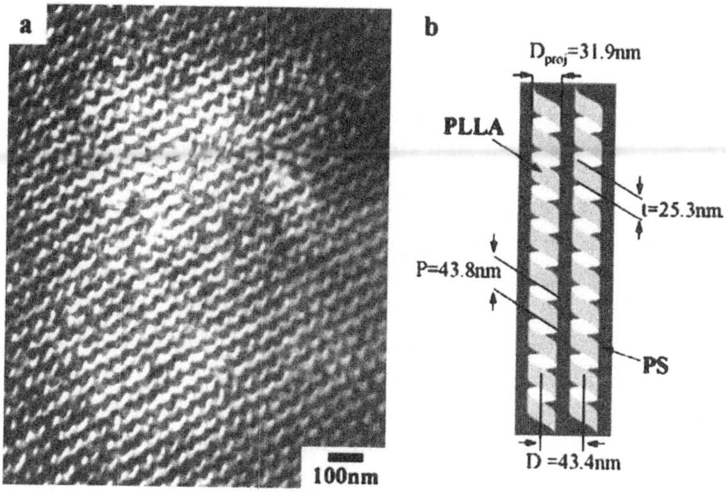

Fig. 3 **a** TEM micrograph of polystyrene-*b*-poly-(*L*-lactide), PS-*b*-PLLA, neat (ϕ_{PLLA} = 0.35). **b** Schematic representation of nanohelical morphology. From [16]. Copyright 2004 American Chemical Society

to persist even after hydrolysis leading to well-ordered nanohelical channels [16].

2.1 Phase Diagrams of Various Block Copolymer Systems: Theory

The phase behaviour of block copolymers can be described theoretically by different approaches [3, 4]. According to mean-field theory (MFT) [17], the phase behaviour of diblock copolymers is dictated by the Flory–Huggins segment–segment interaction parameter, χ, the degree of polymerization, N, and the composition, ϕ, with the product $N\chi$ determining the degree of segregation. In the case $N\chi \leq 10$, the system is ruled by entropic terms, resulting in a disordered phase. Under the condition $N\chi > 10$, enthalpic terms dominate, causing an order-disorder transition (ODT) where the unlike segments segregate into a variety of ordered periodic microstructures. Leibler [17] was the first to describe microphase separation, i.e. the transition between a disordered liquid or homogeneous state, and an ordered one. He treated the microphase separation transition within the so-called weak segregation limit (WSL). The WSL approximates the composition fluctuations by a single wave function (first harmonic). Using higher degrees of segregation between the microdomains leads to a description called the strong segregation limit (SSL). Here, composition fluctuations are approximated by step functions across the microdomain boundaries. An expression of the free energies for classic morphologies in the SSL was first developed by Meier [18] and later on by

Semenov [19], to mention just a few of several groups in this field. Although providing useful insights, neither the SSL nor the WSL describes the morphological behaviour completely. Matsen used a self-consistent field theory (SCFT) which also takes higher harmonics into account and could mostly bridge the gap between weak and strong segregation limit [20].

Additional factors play an important role in determining the phase state: architecture [21–25], fluctuation effects [26, 27] and conformational asymmetry [28–31]. The architecture, i.e. the way the unlike blocks are connected, was shown to have a strong effect on the phase behaviour by changing both the ODT temperature, T_{ODT}, and the boundary between ordered phases. Fluctuation effects decrease the T_{ODT} and allow for direct transitions between phases, which are not adjacent in the original phase diagram from Leibler. As a main effect conformational asymmetry shifts the order-order transitions towards compositions richer in the segments with the longer statistical length.

The morphological behaviour of binary triblock copolymers (ABA) was elucidated in a subsequent work [32]. Including higher harmonics into his computation, Matsen's SCFT is able to bridge the gap between diblock and symmetrically composed triblock copolymers. An asymmetry parameter, τ, was introduced which relates the number of segments of the shorter, $\tau\phi_A N$, with that of the longer A block, $(1-\tau)\phi_A N$ (i.e. $0 \leq \tau \leq 1/2$; with the limiting

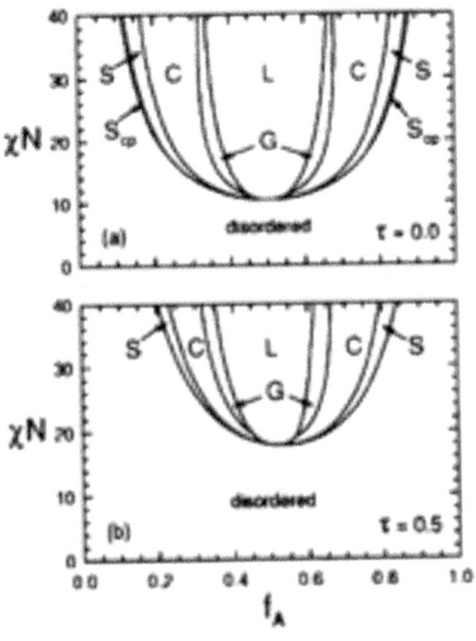

Fig. 4 Mean-field phase diagrams for melts of **a** AB diblock copolymer $\tau = 0$ and **b** symmetric ABA triblock copolymer ($\tau = 0.5$) plotted in terms of segregation χN and composition f_A calculated with SCFT. From [32]. Copyright 2000 American Institute of Physics

cases $\tau = 0$: AB-diblock and $\tau = 1/2$: symmetric ABA triblock). The resulting phase diagrams are shown in Figs. 4 and 5.

Although being qualitatively in agreement with experimental results, disagreements between experiment and theory remain. Besides the composition, f_A, and the total degree of polymerization, N, all theoretical works refer to the segmental interaction parameter χ. This parameter can be estimated from a relationship to the solubility parameters. The ODT as a thermodynamic measure of the incompatibility was used to compare a set of symmetrically composed diblock copolymers from different hydrocarbons, polydimethylsiloxane and poly(ethylene oxide) (PEO) [33]. While the behaviour of hydrocarbon diblock copolymers was successfully described by a consistent set of solubility parameters, this procedure failed for systems containing PEO. The

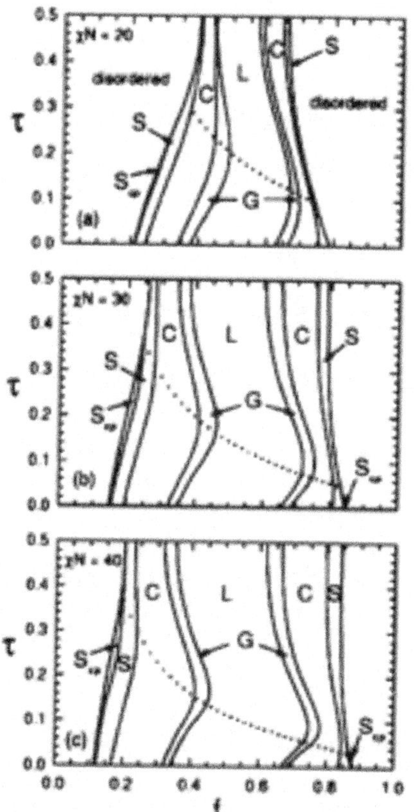

Fig. 5 Mean-field phase diagrams for asymmetric *ABA* triblocks spanning between diblock ($\tau = 0.0$) and symmetric triblock ($\tau = 0.5$) limits at segregations of **a** $\chi N = 20$, **b** $\chi N = 30$ and **c** $\chi N = 40$ calculated with SCFT. *Dotted curves*: critical asymmetries τ_c predicted by SST beyond which short *A* blocks are extracted from their domains. From [32]. Copyright 2000 American Institute of Physics

asymmetry in size of the different components was attributed to an enthalpic contribution to the χ-parameter.

More complex block copolymers as triblock terpolymers have also been described. Molecular dynamics simulations were applied to the morphological behaviour of triblock terpolymers, and basically the experimentally found structures were reproduced [34–36]. Using SCFT, the morphological behaviour of linear triblock terpolymers was described in two dimensions. Depending on composition and segmental interactions, most morphologies were found to be stable as they were described previously in experimental studies [37]. A fundamental problem in most simulation works is boundary conditions like the size and geometry of the simulation box. This can lead the simulation to a wrong equilibrium state. Using a rather flexible simulation approach, Bohbot-Raviv and Wang [38] were able to create morphologies of terpolymers of various topologies in two dimensions without the initial assumptions of the morphology symmetry.

2.2 Phase Diagrams of Various Block Copolymer Systems: Experimental Results

2.2.1 Poly(ethyl ethylene)-*b*-polyisoprene

Chemical modification of one block without changing chain flexibility allows the effective variation of the χ-parameter in a wide range. This was achieved by adding difluorocarbene to the double bonds in poly(1,4-isoprene) blocks of poly(ethyl ethylene)-*b*-poly(1,4-isoprene) (PEE-*b*-PI) copolymers. As a surprising result the gyroid phase was found to be stable even in the SSL, which is in disagreement with the predictions of Matsen's SCFT. The stability of the gyroid phase in the SSL regime was confirmed by annealing of solution cast films. Depending on the solvent, the cast films exhibited lamellar or cylindrical morphologies. In all cases the gyroid morphology was obtained after annealing [39, 40]. However, the unmodified PEE-*b*-PI was only obtained in a lamellar arrangement [41].

2.2.2 Polystyrene-*b*-polyisoprene and Its Multiblock Copolymers

Park et al. [42] constructed a phase diagram spanning a compositional range of about 10% using only a single polystyrene-*block*-polyisoprene, PS-*b*-PI, sample. The test specimen was fractionated twice using normal and reverse phase HPLC techniques. The different fractions exhibited *L*, *PL*, *G* and *C* morphologies after annealing (Table 2). In comparison to unfractionated samples

Table 2 Molecular Characteristics and Phase Behaviour of PS-*b*-PI Block Copolymers of Park et al.'s Study [42]

Mother sample: PS-*b*-PI: $M_n = 34.0\,\mathrm{kg\,mol^{-1}}$; $\phi_{PI} = 0.665$; PL –(170 °C) → G –(230 °C) → Dis				
normal HPLC				
Reverse HPLC	$M_n/\mathrm{kg\,mol^{-1}}$	33.2	34.8	35.0
	ϕ_{PI}	0.706	0.687	0.672
	Transition	C –(200 °C) → Dis	G –(170 °C) → C –(210 °C) → Dis	G –(205 °C) → C –(230 °C) → Dis
	$M_n/\mathrm{kg\,mol^{-1}}$	30.8	32.7	34.3
	ϕ_{PI}	0.675	0.657	0.637
	Transition	G –(180 °C) → C –(210 °C) → Dis	PL –(165 °C) → G –(220 °C) → Dis	L –(165 °C) → PL –(190 °C) → G –(240 °C) → Dis
	$M_n/\mathrm{kg\,mol^{-1}}$	30.6	31.2	32.4
	ϕ_{PI}	0.638	0.627	0.608
	Transition	PL –(165 °C) → G –(245 °C) → Dis	PL –(190 °C) → G –(230 °C) → Dis	L –(245 °C) → Dis

this technique yielded sharper phase boundaries and better defined morphologies mainly due to narrower distributions.

The effects of architecture and block mass on the morphology and properties of linear multiblock copolymers was investigated by Spontak et al. [43]. They compared two matched series of compositionally symmetric $(PS\text{-}b\text{-}PI)_n$ block copolymers with $1 \leq n \leq 4$, one possessing a constant chain mass and the other one having a nearly constant block mass.

The microdomain period depends on the selectivity of the casting solvent (see also Sect. 4.4.2). This effect is most pronounced for the least constrained block copolymer, i.e. the diblock ($n = 1$). The microdomain long period likewise decreases with increasing n in both series. Polymers with $n > 1$ possess at least one PS and one PI midblock. Extension of such midblocks along the lamellar normal is anticipated to be more restricted and in consequence entropically less favoured than that of a single grafted end block. The variances in restriction may account for the differences in casting behaviour and microdomain spacing.

An interesting experiment regarding the lamellae orientation of block copolymers (Fig. 6) was performed by Wu et al. [44]. They aligned a symmetric lamellae-forming PS-*b*-PI-*b*-PS-*b*-PI-*b*-PS-*b*-PI-*b*-PS heptablock copolymer by shearing and characterized the state of microdomain alignment by SAXS and TEM after exposing the copolymer to tetradecane (Fig. 7). As tetradecane is a solvent for PI and a nonsolvent for PS, it will not affect the lamellar structure in case of a bridged chain arrangement. In case of a looped chain arrangement, however, the swelling of the PI domains results in an extensive buckling and delamination. The experimental findings were highly dependent on the processing parameters. For high-strain amplitudes applied at a low shear frequency, a predominantly parallel alignment was

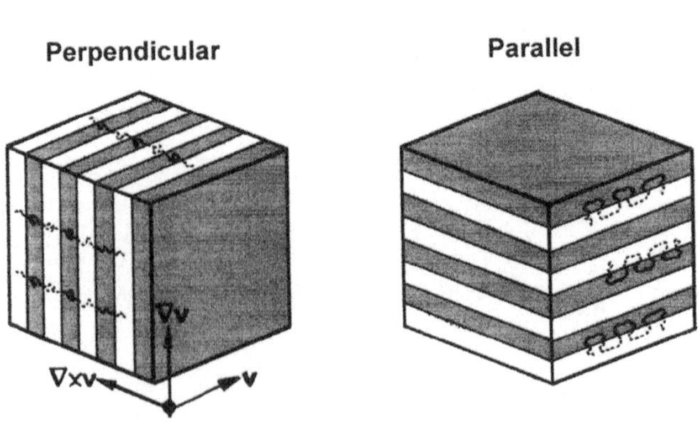

Fig. 6 Idealized heptablock copolymer molecular conformations for parallel and perpendicular orientations. The latter contains predominantly bridged chains, whereas the block copolymer is much more likely to loop along interfaces in parallel arrangement. From [44]. Copyright 2004 American Chemical Society

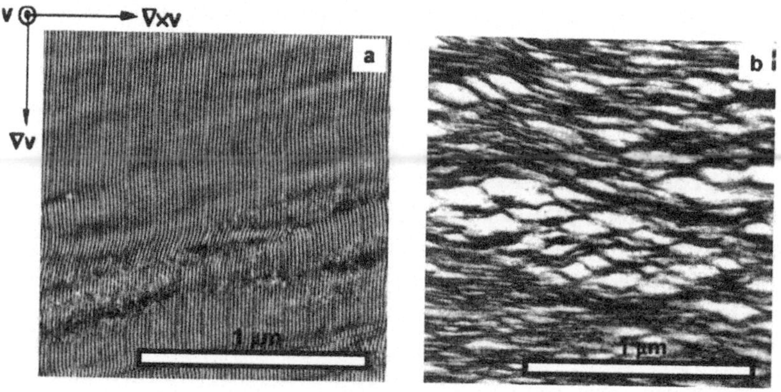

Fig. 7 Representative cryo-TEM micrographs obtained after exposing **a** perpendicular and **b** parallel aligned PS-*b*-PI-*b*-PS-*b*-PI-*b*-PS-*b*-PI-*b*-PS to tetradecane, a solvent for PI and a nonsolvent for PS. PI domains were stained by OsO_4 and appear *dark*. The structure is viewed along shear direction. Perpendicular lamellae were unaffected by solvent, whereas parallel lamellae displayed extensive buckling and delamination, indicating loop formation. The coordinate system is defined in Fig. 6. From [44]. Copyright 2004 American Chemical Society

observed. In contrast, when a low-strain amplitude was applied at high frequencies, the lamellae were mainly perpendicularly aligned. Formation of the latter morphology should favour an extended, bridging-chain conformation depicted in Fig. 6. This arrangement should minimize block stretching during large-amplitude oscillatory shearing. For the parallel morphology shown in Fig. 6, the PI blocks should be largely unconstrained, which implies predominantly looping blocks. This conformation also permits a sliding of layers, consistent with the combination of low shear rate and large strain amplitude. Moreover, these findings suggest a connection between shear-induced microdomain alignment and molecular conformation. The proposed situation is consistent with theoretical models predicting a transformation from bridging to looping and thus generation of sliding layers under large strain amplitudes [45].

2.2.3
Poly(ethylene-*alt*-propylene)-*b*-polydimethylsiloxane

As a model system PS-*b*-PI exhibits a number of disadvantages: The high T_g of polystyrene in conjunction with the thermal instability of PI results in a limited temperature range open for experiments. The chain dynamics of PS is often slow and sluggish, leading to uncertainties in the interpretation of kinetic data.

The drawbacks can be overcome by employing a block copolymer composed of poly(ethylene-*alt*-propylene), PEP, and polydimethylsiloxane

(PDMS). Both components exhibit low glass transition temperatures, T_g, with relatively rapid chain dynamics combined with a higher thermal stability. For instance the ODT from a *bcc* array of spherical micelles into a disordered state was investigated for a PDMS-*b*-PEP system by Wang et al. [46] using temperature-dependent SAXS and DMTA data. In the disordered micellar state some copolymer chains remain in micellar aggregates, but the micelles themselves adopt a liquid-like packing. Further temperature increase leaves the micelles relatively unchanged, but their number, density and volume fraction decreases steadily. However, the authors were not able to detect a critical micelle temperature, *cmt*, in the experimental temperature range.

The diffusion coefficients of this system were determined for disordered micelles and *bcc* spheres [47]. They were found to be retarded as compared to the disordered state. This retardation is consistent with a hindered diffusion process, $D \sim D_0 \exp(-A\chi N_A)$, with D_0 being the diffusion coefficient in the absence of any interactions (i.e. for $\chi \to 0$), and A is a prefactor of order unity. Hence, the diffusion barrier increases with the enthalpic penalty χN_A, where N_A represents the number of monomers in the "foreign" block. In the simplest description of hindered diffusion, the prefactor A remains constant. This model describes the experimental data poorly as A was found to increase with χN_A [47].

In a subsequent study [48] samples were prepared with three different degrees of macroscopic alignment of *hcp* PDMS cylinders. Diffusion coefficients parallel ($D_{\parallel}$) and perpendicular ($D_{\perp}$) to the cylinder axes were resolved. For a high degree of macroscopic alignment both coefficients were retarded as compared to D_0. In well-aligned cylinders there is a large anisotropy of diffusion with $D_{\parallel} > D_{\perp}$ arising from the different mechanisms of parallel and perpendicular diffusion processes. The latter one followed a hindered diffusion mechanism, with a barrier that is exponentially dependent on χN_A. In contrast, $D_{\parallel}$ crosses over to a block retraction mechanism where the diffusion barrier is exponentially dependent on the ratio of degree of polymerization to the entanglement length of the PEP block, $N_{PEP}/N_{e,PEP}$, but remains independent of χ. Additionally, $D_{\perp}$ was found to be strongly dependent on the degree of alignment, while $D_{\parallel}$ was not. Supposedly, misaligned cylinders as well as dislocations accelerate the diffusion in the perpendicular orientation, which results in a larger apparent $D_{\perp}$.

Small-angle X-ray scattering experiments have been performed on an oriented sample of the lamellar phase of PEP-*b*-PDMS copolymers as a function of temperature. The lamellar spacing was found to increase for decreasing temperature below ODT. This was attributed to an increasing incompatibility between the unlike blocks. As the experiments were performed in a range $14 \le \chi N \le 16$ the system resides in the WSL. However, the relation between lamellar spacing and χ—originally derived for the SSL [19]—was found to be in quantitative agreement with the experimental data. Furthermore, a broadening of the Bragg peak was observed. Quantitative determination of the broadening revealed that this was due to the splay deformation and compressional elasticity of the lamellar layers. Near the ODT the amplitude of the diffuse scattering increases, indicative of a decrease in elastic moduli. Due to orientational misalignments in the structure, a quantitative calculation of the layer compressibility modulus was not performed [49].

2.2.4
Polyisoprene-*b*-poly(ethylene oxide)

The phase behaviour of the system polyisoprene-*b*-poly(ethylene oxide), PI-*b*-PEO, has been studied extensively [50–52]. The phase behaviour is governed by the high interaction parameter and the strong tendency of PEO to crystallize (Fig. 8). The phase space consists of five ordered phases: crystalline lamellar (*CL*), amorphous lamellar (*L*), hexagonal array *hcp* of cylinders (*C*), *bcc* array of spheres (*S*) and a gyroid phase (*G*). The *CL* phase dominates the phase behaviour at intermediate and strong segregations. The *L* phase has a peculiar appearance owing to the *CL* formation at intermediate segregation. The *G* phase is found on both sides of the phase diagram, being the last before disordering. For some volume fractions the *C* phase has direct access to the disordered state. Due to the chosen compositions, the *S* phase was only found on one side of the phase diagram. The phase diagram allows for direct transitions between the *C* and *G* phases and between the *L* and *G* phases.

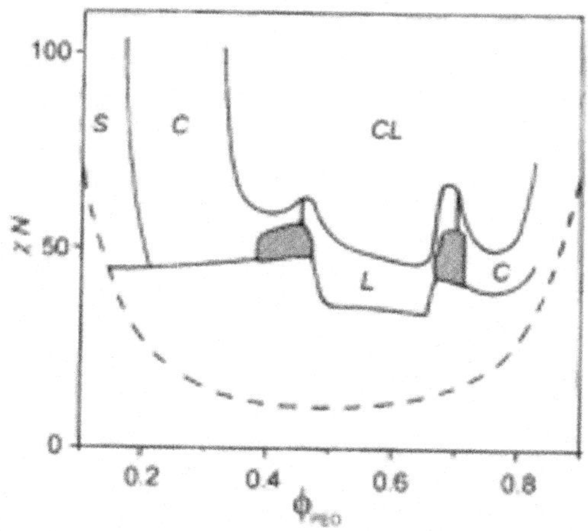

Fig. 8 Phase diagram for PI-*b*-PEO system. Only equilibrium phases are shown, which are obtained on cooling from high temperatures. ODT and OOT temperatures were identified by SAXS and rheology. Values of χN were obtained using $\chi N = 65/T + 0.125$. *Dashed line*: spinodal line in mean-field prediction. Note the pronounced asymmetry of phase diagram with ordered phases shifted parallel to composition axis. Asymmetric appearance can be accounted for by conformational asymmetry of segments. Adopted from [53]

2.2.5
Polystyrene-*b*-poly(*n*-pentyl methacrylate)

An unusual phase diagram was found for deuterated polystyrene-*b*-poly(*n*-pentyl methacrylate), *d*PS-*b*-P*n*PMA. With increasing temperature a disorder-order transition takes place (LDOT). At higher temperatures an additional transition from the ordered to the disordered state termed an upper order-to-disorder transition (UODT) showed up, leading to the existence of a closed-loop phase diagram [54–57]. The copolymer is phase seperated within the temperature interval between the LDOT and UODT but is disordered outside of this interval. With increasing molecular weight the LDOT decreases, whereas the UODT increases. As the molecular weight decreases, the size of the closed loop decreases and vanishes finally.

At higher temperatures the temperature dependence of χ achieves a maximum. By analogy, below LDOT, χ does not decrease monotonically with decreasing temperature but exhibits a minimum. This strongly suggests that at lower temperatures another ODT might be expected as sketched in Fig. 9.

Regardless of block-length ratios (as long as the molecular weight and ϕ_{dPS} are properly chosen) LDOT and UODT can be observed. Within the two transitions, the ordered microdomains change from *L* to *C* and finally to *S* microdomains, with decreasing ϕ_{dPS} from 0.5 to 0.23. The ordering and disordering upon increasing temperature indicate entropic reasons for this unusual phase behaviour [57].

Interestingly, the phase behaviour of the corresponding blend consisting of *d*PS and P*n*PMAS is entirely different. Closed-loop behaviour was not ob-

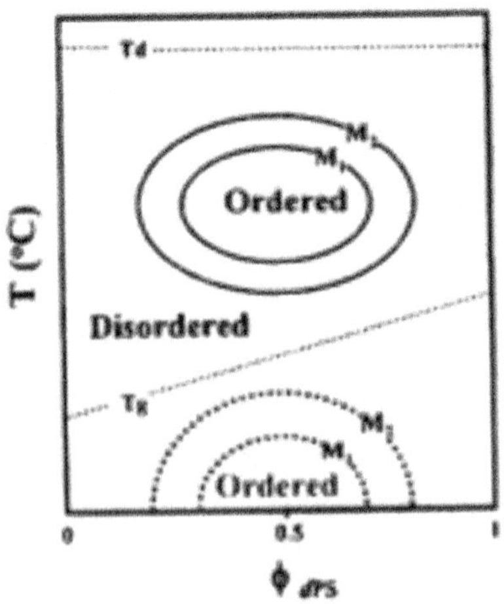

Fig. 9 Schematic of phase behaviour for *d*PS-*b*-P*n*PMA with two different molecular weights ($M_2 > M_1$). T_g and T_d are glass transition and degradation temperatures respectively. From [55]. Copyright 2003 American Chemical Society

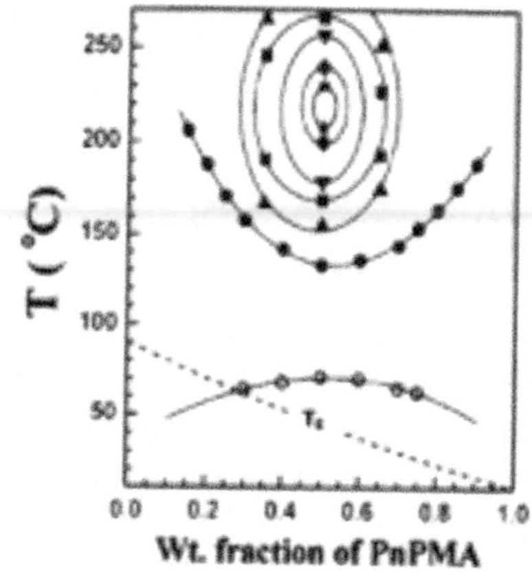

Fig. 10 Phase diagrams of dPS/PnPMA blends. ○ UCST at ambient pressure, • LCST at ambient pressure. Closed-loop phase behaviours are observed at higher pressure: ▲: 97 bar; ■: 117 bar; ▼: 138 bar; ♦: 166 bar; ★: 186 bar. *Dashed line*: Prediction of T_g, blend by Fox equation at ambient pressure. From [59]. Copyright 2004 American Chemical Society

served; however, it might be expected when the molecular weights of PS (or PnPMA) become smaller [58]. No UCST at higher temperatures was detected because of sample degradation at temperatures exceeding 270 °C. The differences in the phase behaviour of the block copolymer and the corresponding blend may be due to subtle changes in the entropic contribution to χ.

By increasing pressure the blend system could be forced to exhibit closed-loop phase diagrams. With increasing pressure the loop size decreases and finally vanishes at 200 bar (Fig. 10). As long as the free volume effects are suppressed by external pressure, a blend exhibiting an LCST tends to become homogeneous at higher temperatures. Translational (or combinatorial) entropic contributions prevail over the combination of the repulsive interactions and the negative entropic contributions due to the free volume. These findings are corroborated by the temperature dependence of the segmental interaction parameter χ_{eff}. With increasing temperature it exhibits a minimum followed by a maximum and finally by a continuous decrease [59].

The blend's phase behaviour can be modified by the presence of low amounts (5 wt %) of symmetric *d*PS-*b*-PnPMA as well as PS-*b*-PnPMA when using the non-deuterated PS for blending. As the block copolymer serves as a compatibilizer for the blend, it increases the LCST but decreases the UCST, leading to an expansion of the miscibility window. However, if the critical amount of symmetric block copolymer is exceeded, the LCST remains constant regardless of blend compositions. Thus, the LCST of asymmetric blend compositions with a symmetric *d*PS-*b*-PnPMA (or PS-*b*-PnPMA) was lower than that for another blend without the block copolymer, suggesting that the block copolymer acts as an incompatibilizer for asymmetric blend compositions [60].

2.2.6
Poly(ethylene oxide)-*b*-poly(1,2-butylene oxide)

The morphologies of various copolymers composed of poly(ethylene oxide), PEO, and poly(1,2-butylene oxide), PBut, were recently reviewed by Ryan et al. [61]. The corresponding phase diagrams of PEO-*b*-PBut, PBut-*b*-PEO-*b*-PBut and PEO-*b*-PBut-*b*-PEO melts are depicted in Fig. 11. In all phase diagrams the semi-crystalline lamellar phase was not attained because of the copolymers' low melting points.

Moving across the PBut-*b*-PEO phase diagram (Fig. 11a) from low to high fractions of PEO the phases are (i) *S*; (ii) *C*; (iii) *G*, bounded at low temperature by either *C* or *PL*; (iv) *L*; (v) *G*, bounded at higher χN by *L* or transient *PL* (see below and cf. Fig. 2) and (vi) *C*. A second *S* region for high ϕ_{PEO} was not located in the examined concentration range. The PBut-*b*-PEO-*b*-PBut copolymer (Fig. 11b) exhibits features similar to those of the diblock. Both have a similar *G* region, and the transition between the disordered and *G* phases oc-

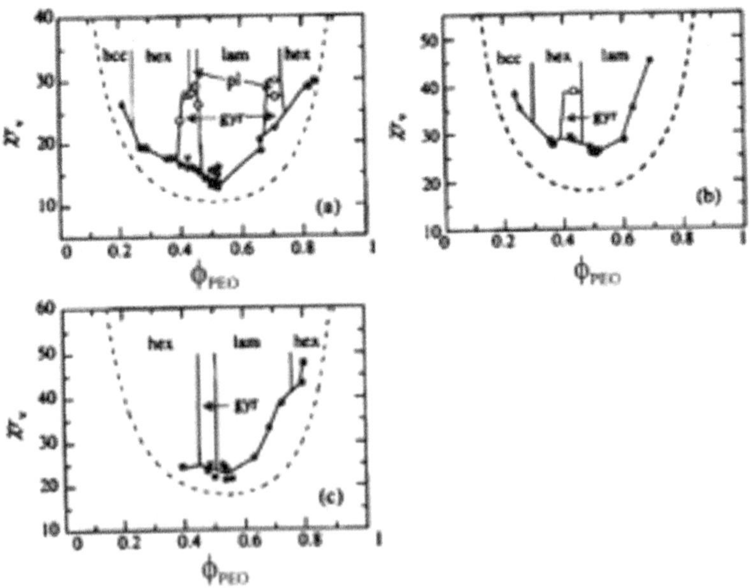

Fig. 11 Phase diagrams for **a** PBut-*b*-PEO, **b** PBut-*b*-PEO-*b*-PBut, and **c** PEO-*b*-PBut-*b*-PEO. Values of χ were calculated for fluctuating melts and appropriate transition temperature. •, ■, ▲, ▼: ODTs; ○, □, ◇: OOTs. •, ○: *L*; ■, □: *C*; ▲: *S*; ▼: *G*; ◇: *PL*. *Dashed lines*: boundaries predicted by exact SCFT with $\chi = 0.6$. The *solid curves* are freely drawn to indicate various phases. From [61]. Copyright 2001 Owner Societies

curs at higher χN than expected for a smooth boundary between the *L* and *C* regions. The PEO-*b*-PBut-*b*-PEO phase diagram (Fig. 11c) was only accessible for $\phi_{PEO} > 0.5$; despite the limited range, the similarity between the results for the three is obvious. Similarities between Fig. 11 and phase diagrams of PS-*b*-PI [11] and PEP-*b*-PDMS [13] are also worth while noticing. All three systems include *G* and *PL* regions. Generally, higher molar masses can be obtained in the PS-*b*-PI system leading to higher accessible values of χN. The system reveals a significant dependence of the domain spacing of the *L* phase on the block architecture. Triblock copolymers are 10 to 20% more stretched than their diblock analogues, which is in good agreement with MFT predictions. The more extended conformation of short triblocks results in the ability of their centre blocks to bridge the lamellar domains [62].

2.2.7 Poly(ethylene-*alt*-propylene)-*b*-poly(*d*,*l*-lactide)

The phase space of the poly(ethylene-*alt*-propylene)-*b*-poly(*d*, *l*-lactide) (PEP-*b*-PLA) system was recently explored by Schmidt and Hillmyer [63]. All equilibrium symmetries *L*, *C*, *S*, and *G* were observed. The *G* morphology, however, existed in only a small region between the *L* and *C* in close proximity to the ODT (Fig. 12).

The experimental morphology diagram, Fig. 12, is rather symmetric as observed by the distinct lamellar region containing the boundaries between

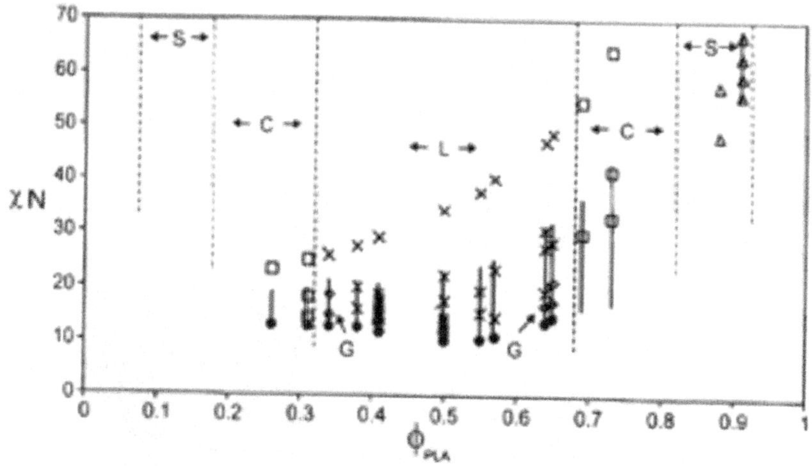

Fig. 12 Morphology diagram of PEP-*b*-PLA. •: ODTs determined by rheology; △, □, ◇, ×: ordered microstructures directly observed by SAXS; △: *S*; □: *C*; ◇: *G*; ×: *L*. *Solid lines*: ordered range of χN as determined by rheological measurements; *dashed lines*: approximate phase-transition boundaries using experimental data and mean-field theory predictions. From [63]. Copyright 2002 Wiley

the *L* and *C* region. The interaction parameter of $\chi_{PEP\text{-}PLA}(140\,^{\circ}C) = 0.44$ is relatively large; hence, short-chain lengths are required to access an experimentally viable range of ODTs. Despite the low molecular weight species used for this investigation, deviations from the morphological behaviour of long-chain diblock copolymer systems were not observed, which is also consistent with theoretical predictions.

2.2.8
Polystyrene-*b*-poly(2-vinylpyridine)-*b*-poly(*tert*-butyl methacrylate)

A systematic comparative study of triblock terpolymers in the bulk and thin-film state was carried out on polystyrene-*b*-poly(2-vinyl pyridine)-*b*-poly(*tert*-butyl methacrylate), PS-*b*-P2VP-*b*-P*t*BMA. A diblock precursor with a minority of PS leading to a double gyroid structure was used. Upon increase of P*t*BMA content this morphology changed from lamellae with

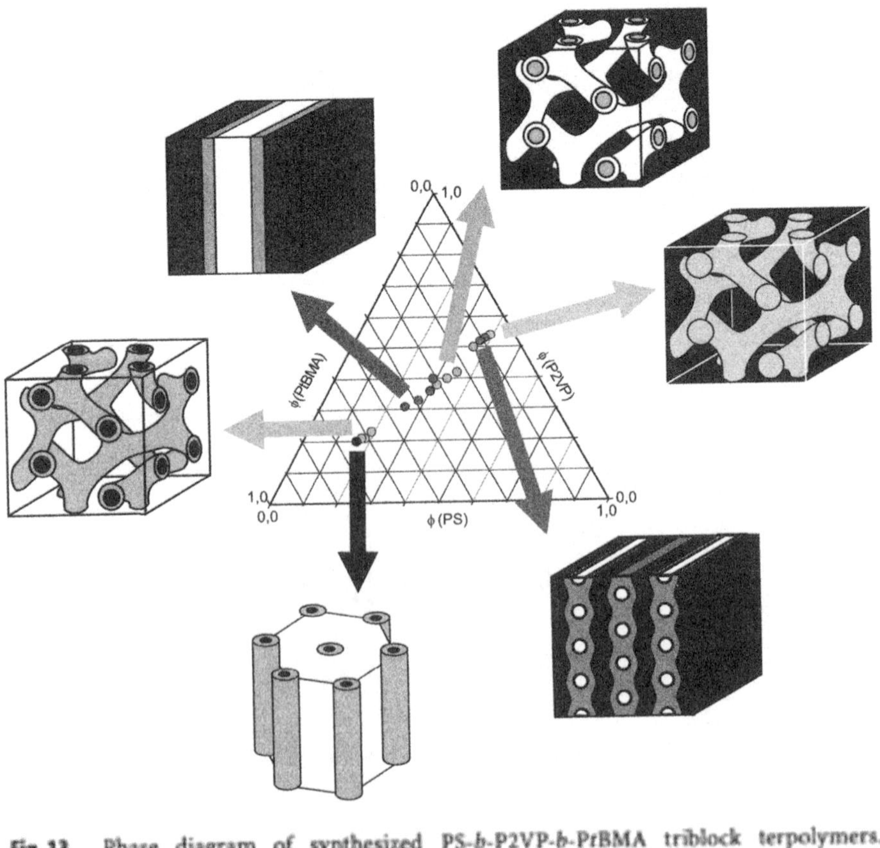

Fig. 13 Phase diagram of synthesized PS-*b*-P2VP-*b*-P*t*BMA triblock terpolymers. From [64]. Copyright 2003 Elsevier

P*t*BMA spheres in P2VP, to core-shell double gyroid with P*t*BMA cores, lamellae, core-shell double gyroid with PS core (in coexistence with perforated lamellae), and finally core-shell cylinders with PS core [64]. The morphology changes are schematized in Fig. 13. One of the systems was also investigated as thin film, and also the corresponding bulk morphology was found (core-shell double gyroid in coexistence with perforated lamellae) [65].

In contrast to diblock copolymers, the more complex architecture of terpolymers aggravates the route to thermodynamic equilibrium. The high viscosity of the terpolymers and the rather narrow temperature window between the highest glass transition and the lowest thermal degradation temperature of the blocks does not necessarily allow thermal annealing. This may also result in severe kinetic barriers for complex metastable structures formed during sample preparation processes. The barriers arise from strong segregation powers among different block chains and have to be overcome in order to achieve equilibrium structures. Therefore, swelling the systems with non-selective solvents, followed by slow drying of the systems, proves an effective alternative route for equilibration.

A broad variety of different morphologies depending on the preparation technique can be realized in the PS-*b*-P2VP-*b*-P*t*BMA system. Dip-coating the triblock terpolymer ($\phi_{PS} = 0.17$, $\phi_{P2VP} = 0.26$ and $\phi_{PtBMA} = 0.57$) from chloroform and subsequent solvent annealing with tetrahydrofuran produced a 2D perfect alternating order with the P2VP block winding about the PS core in a left- and right-handed helical pattern as shown in Fig. 14. The helicity does not change along a cylinder and strictly alternates between neighbouring ones. The appearance of that spectacular morphology was attributed to packing arguments. In a 2D environment an alternating helicity of neighbouring cylinders yields a closer packing (56 ± 2 nm) compared to a *hex* arrangement (64 ± 2 nm), a situation which may be envisioned by two screws of opposite chirality fitting in each other, as sketched in Fig. 15. The denser packing may result in a release of unfavourable chain stretching. In bulk, however, a strict alternation between nearest neighbours is impossible in a *hex* arrangement. Hence, the system becomes frustrated and assembles in a monolayer of cylinders composed of a PS core (gray) surrounded by a P2VP shell (bright) embedded in a matrix of P*t*BMA (dark) (Fig. 14) [66].

Spin-coating of PS-*b*-P2VP-*b*-P*t*BMA triblock terpolymers ($0.18 \le \phi_{PS} \le 0.20$; $0.39 \ge \phi_{P2VP} \ge 0.34$, and $0.43 \le \phi_{PtBMA} \le 0.46$) on a polar substrate produces films exhibiting a trapped sponge-like microphase separated structure covered by a thin homogeneous layer of P*t*BMA on the surface [67]. Upon annealing with THF vapor, lamellae evolve from the film's free surface and finally will be developed throughout the entire film. In the region near the polar substrate, a thin, laterally inhomogeneous microphase-separated layer is formed. It allows the polar middle block to adsorb at the polar substrate. After prolonged treatment, the non-equilibrium lamel-

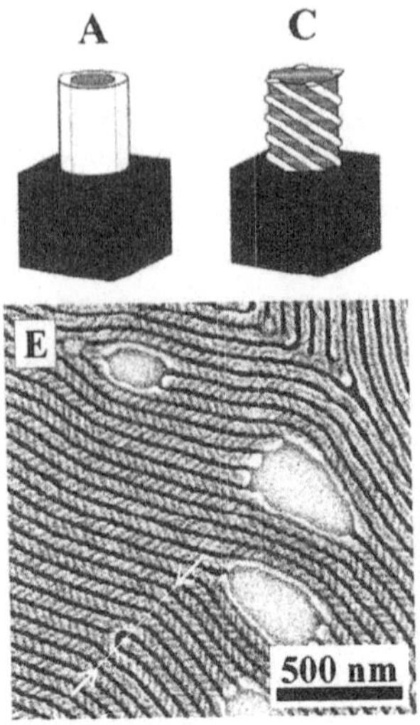

Fig. 14 Sketch of ABC triblock terpolymer morphologies: CSCs (**a**) and helices wound around cylindrical core (**c**) SEM images of 60-nm-thick PS-b-P2VP-b-PtBMA film after staining with OsO_4 (**e**). *Bright* structures are expected to be P2VP blocks, *grey* ones should be PS ones, while PtBMA blocks appear *darkest*. *White line*: change in helicity along single cylinders induced by line defects. From [66]. Copyright 2001 American Chemical Society

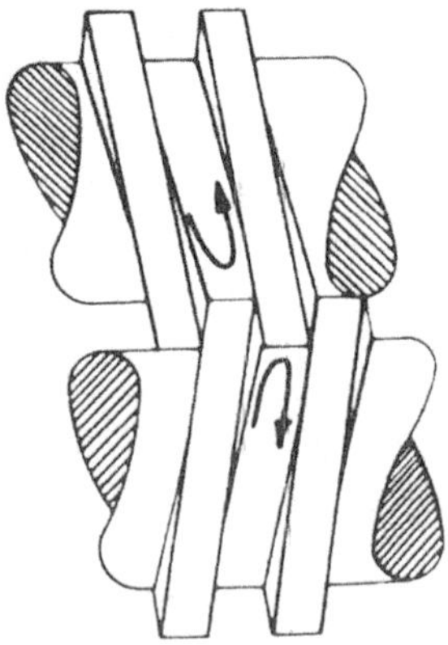

Fig. 15 Sketch of two screws with opposite chirality fitting tightly into each other. From [70]. Copyright 2002 American Chemical Society

lar microdomains are annealed to thicker equilibrated ones accompanied by macroscopic dewetting of the film from the substrate. The process is sketched in Fig. 16.

In a subsequent publication [68] the influence of the substrate on the self-assembly process was elucidated. The comparison of a polyimide substrate with a native oxide (SiO_x) covered wafer revealed that differences in interac-

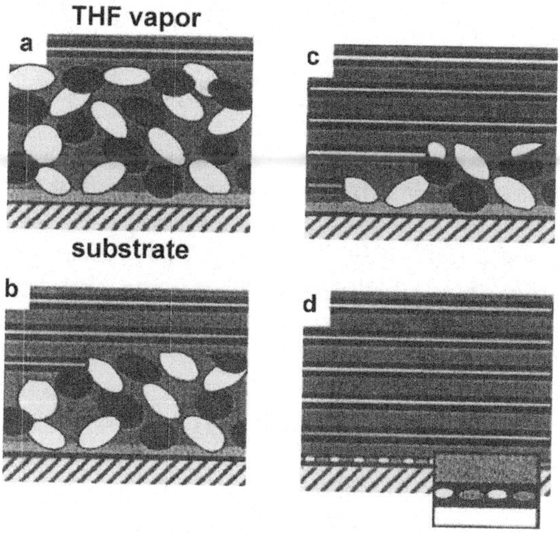

Fig. 16 Schematic diagram of microdomain self-assembly in thin PS-*b*-P2VP-*b*-P*t*BMA triblock terpolymer films. The prepared film exhibits sponge-like distorted microdomain structures in bulk with homogeneous layer of P*t*BMA on free surface. Homogeneous layer triggers surface-induced organization into three-phase coexisting lamellae. Upon THF vapor treatment, lamellar grains evolve from free surface and grow into bulk of film (**b**, **c**). Finally, lamellae with uniform multilayered structure throughout the film (**d**) are formed. Strong interactions of middle block to substrate results in a laterally microphase-separated structure (inset of (**d**)) formed in region adjacent to substrate. From [67]. Copyright 2002 American Chemical Society

tion strength between substrate and polymer cause different configurations of the polar P2VP middle block. As a consequence, the structure of the first layer can be altered by the choice of the substrate which influences the self-assembly of the lamellae throughout the entire film.

Interestingly, when spin-coating from chloroform solution, the packed cylinder structure of PS-*b*-P2VP-*b*-P*t*BMA (cf. Fig. 10, $\phi_{PS} = 0.16$, $\phi_{P2VP} = 0.21$ and $\phi_{PtBMA} = 0.63$) will self-assemble in a perforated lamellar structure [69]. This morphology may be visualized as a P2VP/PS/P2VP sheet perforated by P*t*BMA channels, which themselves connect between two outer P*t*BMA layers. The bicontinuous structure is remarkably stable. Neither the removal of the P*t*BMA component by ultraviolet radiation nor saponification of the *tert*-butyl groups altered the structural motif, indicating potential applications for stimulus-sensitive membranes and for lithographic applications. The differences in morphology between the thin film and the bulk material indicate that the near-surface structure of a thick (bulk) sample can deviate considerably from the inner parts of the material.

Moreover, the solvent-annealing process is rather complex and the underlying delicate balances of enthalpic and entropic contributions to the free energy are not yet understood in detail. Besides the solvent type, the solvent extraction rate also has an influence on the final surface structure [70]; however a detailed discussion of the effects is beyond the scope of this contribution, and the interested reader is referred e.g. to [70–72].

2.2.9 Poly(ethylene-*alt*-propylene)-*b*-poly(ethylene-*co*-butylene)-*b*-polystyrene

A reversible succession of order-order and order-disorder transition was observed for a poly(ethylene-*alt*-propylene)-*b*-poly(ethylene-*co*-butylene)-*b*-polystyrene terpolymer, which shows at room temperature non-hexagonally packed PS cylinders. Upon heating, this system reorganizes to a hexagonally packed one, and at higher temperatures dynamic-mechanical analysis indicates the transition to the disordered state [73].

2.2.10 Polyisoprene-*b*-polystyrene-*b*-poly(vinyl methyl ether)

A triblock terpolymer consisting of polyisoprene, deuterated PS and poly(vinyl methyl ether) shows a UCST behaviour between the first two blocks and an LCST behaviour between the last two blocks. Although the UCST was not

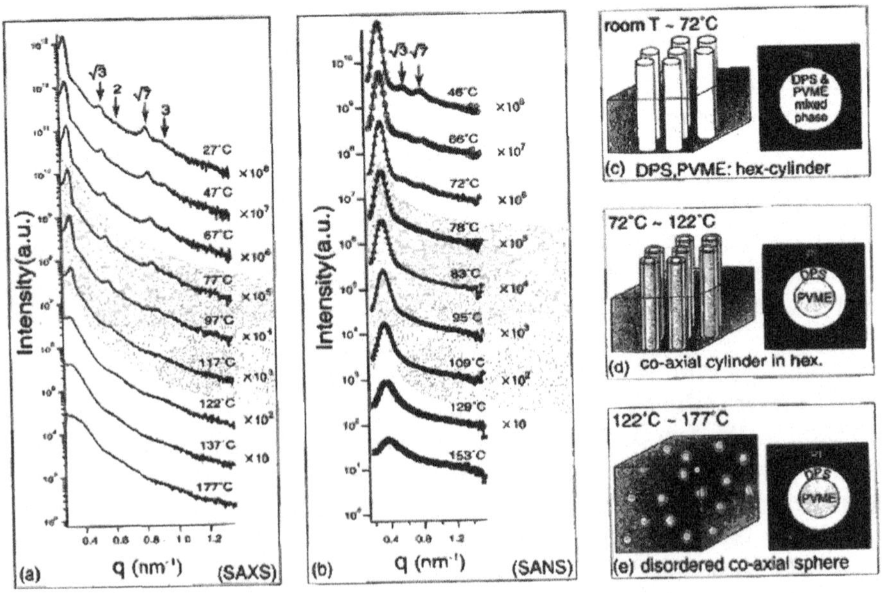

Fig. 17 In situ SAXS (**a**) and SANS profiles (**b**) obtained at various temperatures and schematic models of microdomain structures in temperature ranges between room temperature and 72 °C (**c**), between 72 and 122 °C (**d**), and between 122 and 177 °C (**e**). From [74]. Copyright 2003 Wiley

reached, also in this system a succession of several morphological transitions was reported as indicated in Fig. 17 [74].

2.2.11
Polyisoprene-*b*-polystyrene-*b*-poly(ethylene oxide)

The phase diagram depicted in Fig. 18 summarizes Epps et al. [75, 76] results of SAXS, TEM, dynamic mechanical spectroscopy and static birefringence for the polyisoprene-*b*-polystyrene-*b*-poly(ethylene oxide), PI-*b*-PS-*b*-PEO, system. Besides the well-known two-domain and three-domain lamellae morphologies [77], three co-continuous morphologies were observed. Two of them, the core-shell double gyroid (space group: $Ia\bar{3}d$, denoted Q^{230}) and the alternating gyroid (or tricontinuous double gyroid; space group: $I4_132$, denoted Q^{214}) are of cubic symmetry, whereas the third one (space group: *Fddd*, denoted O^{70}) is an orthorhombic network phase. Those symmetries have already been observed in other ternary triblock systems, e.g. the core-shell double gyroid in PI-*b*-PS-*b*-PDMS [78], the alternating gyroid in PI-*b*-PS-*b*-P2VP systems [79–81] and the O^{70} symmetry in poly(cyclohexyl ethylene)-*b*-poly(ethyl ethylene)-*b*-polyethylene [82]; however, a succession of two and more 3D network morphologies in a homologous set of triblock terpolymers has not yet been reported.

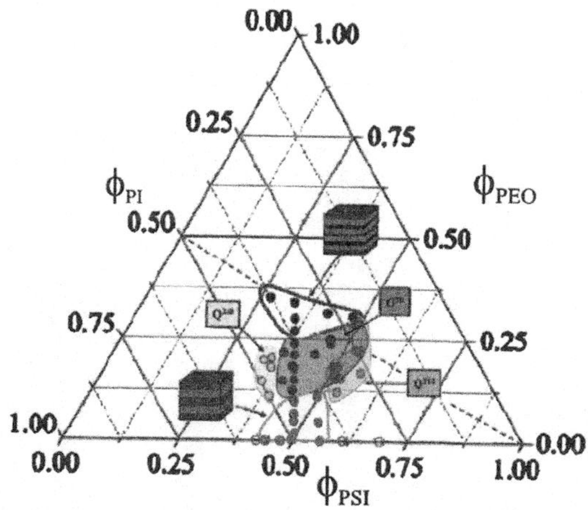

Fig. 18 Phase space of PI-*b*-PS-*b*-PEO in vicinity of ODT. *Filled* and *open circles*: ordered and disordered states, respectively, within experimental temperature range $100 \le T/°C \le 225$. *Outlined areas*: compositions with two- and three-domain lamellae (identified by sketches); *shaded regions*: three network phases, core-shell double gyroid (Q^{230}), orthorhombic (O^{70}), and alternating gyroid (Q^{214}). Overlap of latter two phase boundaries indicates high- and low-temperature occurrence, respectively, of each phase. *Dashed line*: condition $\phi_{PI} = \phi_{PEO}$ associated with symmetric PI-*b*-PS-*b*-PEO molecules. From [75]. Copyright 2004 American Chemical Society

The network phases feature complicated topologies that are difficult to visualize. The various morphologies were modelled using Fourier synthesis techniques. The space-filling models derived from this calculation are shown in Fig. 19a–c for each domain. The composite block copolymer morphology which was obtained by assembly of separate domains into a single structure is shown in Fig. 19d. Based on these real-space models a 2D simulation of the TEM result was performed by integrating the contrast level through a sample of constant volume. The results are shown in Fig. 19e exhibiting an excellent degree of congruity between simulated and experimental TEM data (Fig. 19f, see below). By comparing 2D cuts through the level set models, the nature of the network channels can further be explored. The unit cell of each network structure and a direct projection of each exposed cross section are given in Fig. 20. These cross sections appear to scale; that is, the core-shell double gyroid lattice constant is roughly twice the lattice constants of both the alternating gyroid lattice and the *a* axis of orthorhombic network.

A TEM micrograph of the core-shell double gyroid phase (Q^{230}) is depicted in Fig. 19f, top row. It displays a typical "wagon-wheel" pattern with threefold junctions forming the "spokes" of the wheel. The phase occurs at compositions strictly obeying $\phi_{PI} > \phi_{PS} > \phi_{PEO}$ (Fig. 18). It can be anticipated that this sequencing prefers a "core-shell" interfacial configuration with parallel PI/PS and PS/PEO interfaces enclosing the PEO domain. The lattice consists of two

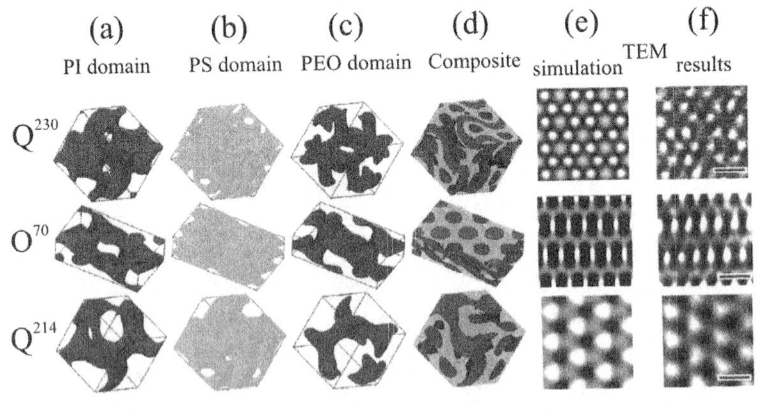

Fig. 19 **a–d**: space-filling domain models of (*upper row*): core-shell double gyroid (Q^{230}), (*middle row*): orthorhombic phase (O^{70}), and (*bottom row*): alternating gyroid phase (Q^{214}). **e** TEM simulation. **f** TEM results. Images were constructed from isosurfaces of a triply periodic field generated by Fourier synthesis to have correct space group symmetry. **a** PI domain, **b** PS domain, and **c** PEO domain; fully assembled structure is shown in (**d**). Greyscale shading and projection to two dimensions ([111] and [110] directions for cubic and orthorhombic systems respectively) connects SAXS data and TEM images. PI-*b*-PS-*b*-PEO specimens were stained with osmium tetraoxide. Scale bars in TEM images represent 30 nm. Adopted from [75, 76]

enantiomeric nets, each of which represents a PEO-rich region surrounded by the PS domain, with PI as the matrix component (Figs. 19a and 20, first column). The two networks composing the centrosymmetric morphology lead to the formation of two distinct PEO and PS domains.

The TEM micrograph of the alternating gyroid phase (Q^{214}) reproduced in Fig. 19f (bottom row) highlights the threefold symmetry projection, which can be observed along with a fourfold one (not reproduced here). The prevalence of threefold and fourfold symmetry in micrographs is consistent with the cubic space group assignment. The phase is located on the PS-rich side of the phase space (Fig. 18), which is roughly consistent with the results of SCFT, which predicts its occurrence when $\phi_A = \phi_C$ and $\chi_{AB} = \chi_{BC} < \chi_{AC}$ [83]. The alternating gyroid persists well into the PEO-lean side of the $\phi_{PI} = \phi_{PEO}$ isopleth, indicating either a large region of stability in general or a shift of phase boundaries due to conformational asymmetry. The middle block PS domain accounts for the greatest volume fraction. Consequently, the PI/PS and PS/PEO interfaces are oppositely curved.

From Fig. 19 the topological equivalence of the alternating gyroid phase to the core-shell double gyroid one is obvious. Both lattices feature two enantiomeric nets. However, in the alternating gyroid each net is chemically distinct, i.e. the PI block occupies one net while the PEO block resides in the other (Fig. 19a and c, or Fig. 20, right column), leading to a chiral, non-centrosymmetric structure. Because the enantiomeric nets are spanned by

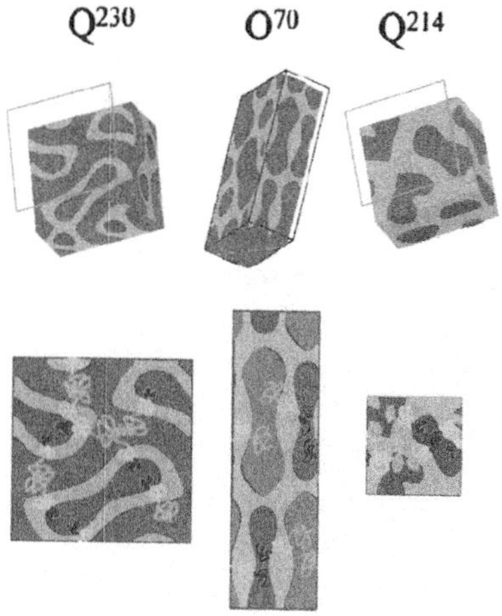

Fig. 20 *Top row*: single unit-cell models of core-shell double gyroid (Q^{230}), orthorhombic (O^{70}), and alternating gyroid (Q^{214}) cross-sectioned to reveal interfacial configuration. *Bottom row*: direct projections of cross-sectioned interfaces. Sketches of PI-*b*-PS-*b*-PEO chains show how each morphology is assembled. Projections appear to scale; that is, the core-shell double gyroid unit cell is roughly twice the thickness of the other two. From [75]. Copyright 2004 American Chemical Society

a single chain, it can be intuitively understood that the lattice dimension of the alternating gyroid is roughly half that of the double gyroid counterpart, whose nets are bridged by two polymer chains.

The TEM micrographs (a representative example is depicted in Fig. 19f, middle row) of the orthorhombic network phase (O^{70}) exhibit no discernible contrast between the bright PS and PEO domains, whereas the OsO_4-stained PI one appear black. The pattern consists of staggered rows of bright elliptical spots interconnected with an array of trivalent connections. The domain connectivity suggests that the morphology exhibits a triply periodic continuously percolating network structure. However, the micrographs lack regions of true threefold and fourfold symmetry common to cubic structures. The large birefringence signal of those materials is a consequence of the optical anisotropy of the morphology. In summary these data confirm the presence of a noncubic symmetry, since cubic structures are optically isotropic over mesoscopic-length scales.

The orthorhombic network phase occupies the largest region of the phase space (Fig. 18). Here, the PI and PS domains become comparable in size, and the degree of curvature required to assemble the PI block into a gyroid topology presumably incurs a destabilizing elastic penalty. The orthorhombic morphology best accommodates these intermediate compositions, allowing the optimization of elastic energy and interfacial tension by providing varying degrees of interfacial curvature. Level set models show that the lattice is based on the *Fddd* symmetry and that the basis consists of segregated triblock terpolymers. This morphology can be envisioned as containing two interpenetrating nets, related by translation, forming the disproportionately sized PI and PEO domains (Fig. 20, middle column), separated by a "matrix" of PS. Both networks are constructed from planar, trivalent connectors assembled into ten-node loops.

2.2.12
Influence of Block Sequence on Morphology

The influence of the block sequence on the morphology of triblock terpolymers has been found to be significant as there are different interaction parameters between the different blocks. Pronounced influence of the block sequence was reported for the PS-*b*-PI-*b*-P2VP and PI-*b*-PS-*b*-P2VP systems with 1 : 1 : 1 composition. Due to the asymmetric interactions ($\chi_{PI\text{-}P2VP} \gg \chi_{PI\text{-}PS} \approx \chi_{PS\text{-}P2VP}$), the former system forms core-shell cylinders with P2VP core, PI shell and PS matrix [84], while the latter forms lamellae [85]. A similar behaviour was found in PS-*b*-PB-*b*-P2VP and PB-*b*-PS-*b*-P2VP triblock terpolymers. For PS-*b*-PB-*b*-P2VP and $\phi_{PS} \approx \phi_{PB}$, core-shell cylinders were observed for $\phi_{P2VP} \approx 0.1$. With increasing amounts of P2VP core-shell double gyroids with P2VP cores and lamellae were found, whereas the other series showed only lamellae [86].

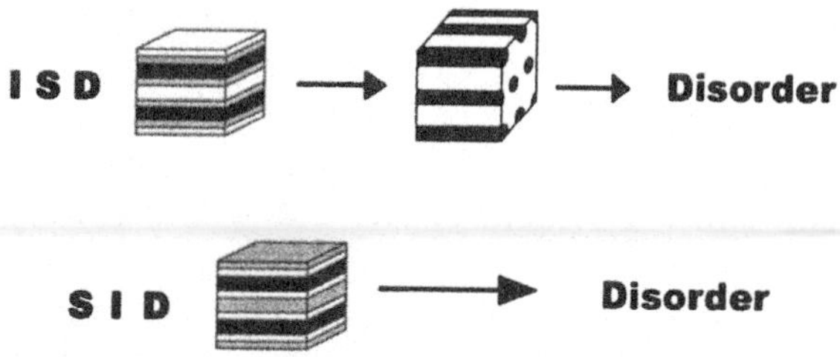

Fig. 21 Schematic summary of thermodynamic phase behaviour for neat, compositionally symmetric PI-*b*-PS-*b*-PDMS (ISD) and PS-*b*-PI-*b*-PDMS (SID) terpolymers. Heating PI-*b*-PS-*b*-PDMS or decreasing molecular weight causes transition from three-domain lamellae to hexagonally packed two-domain cylindrical morphologies, followed by disordering. PS-*b*-PI-*b*-PDMS disorders directly from three-domain lamellar state. From [88]. Copyright 2002 American Chemical Society

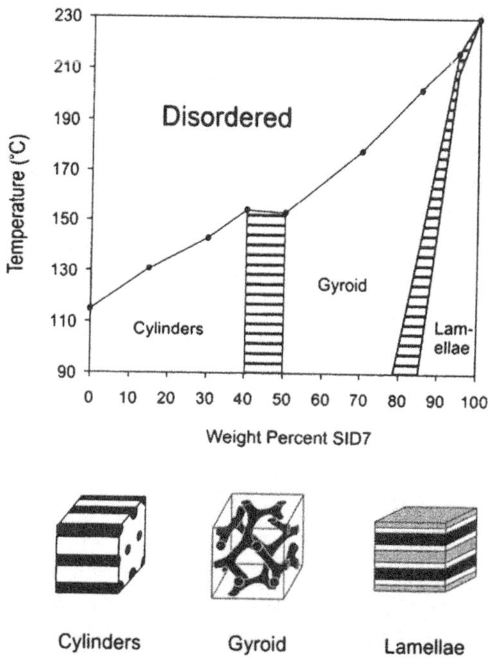

Fig. 22 Phase diagram for blends of PS-*b*-PI-*b*-PDMS (SID7) and PI-*b*-PS-*b*-PDMS. *Horizontal patterned lines* in plot: coarsely defined coexistence regions. Increasing composition of SID7 causes a progression from cylindrical (two-domain), via gyroid (two-domain), to lamellar (three-domain) morphologies. These blend samples exhibit classical alloying behaviour throughout composition range, generally achieving long-range order below T_{ODT} through mixing of *S* and *I* blocks. From [88]. Copyright 2002 American Chemical Society

Although the phase space of the PI-*b*-PS-*b*-PEO system has been extensively discussed in the preceding chapter, it is instructive to compare it with a system in which the block sequence of the first blocks is inverted. Unlike the results of the PI-*b*-PS-*b*-PEO system, where for the condition $\phi_{PI} = \phi_{PS}$ the two lamellar structures and the orthorhombic O^{70} structure were detected (cf. Fig. 18 and the isopleth phase diagram depicted in Fig. 55a in Sect. 4.4.2) the phase space of the PS-*b*-PI-*b*-PEO exhibited lamellae, a tentatively assigned pillared lamellar structure, hexagonal core-shell cylinders, core-shell gyroids, three-domain semiperforated lamellae and lamellae of all three blocks [87]. Both isopleth phase diagrams and sketches of the different morphologies are shown in Figs. 54a and 55a. The differences in phase behaviour were attributed to differences in chain packing. Because of unfavourable PI/PEO contacts, the PS-*b*-PI-*b*-PEO system exhibits a higher degree of frustration, which is eliminated in the PI-*b*-PS-*b*-PEO system.

The influence of the block sequence on the ODT was investigated for symmetrically composed lamellar PS-*b*-PI-*b*-PDMS and core-shell cylindrical PI-*b*-PS-*b*-PDMS with PDMS cores of relatively low molecular weight (10 k). The disordering process of the cylindrical structure takes place at a much lower temperature as compared to the lamellar structure, as sketched in Fig. 21. Blends of these two triblock terpolymers show a gyroid structure over a large composition range (Fig. 22). This indicates a reduction of the segmental contacts with the highest energy (between PS and PDMS) [88].

2.2.13
Tetrablock Terpolymers

Recently, Takano et al. [89] published studies on the morphological behaviour of ternary PI-*b*-PS-*b*-P2VP-*b*-PI tetrablock terpolymer of unequal volume fractions ($\phi_{PI,1} = \phi_{PI,2} = 0.23$, $\phi_{PS} = 0.24$ and $\phi_{P2VP} = 0.30$). The self-assembly into three lamellar microphases of the tetrablock system can be realized in two different ways, as sketched in Fig. 23a,c. In the case of a one-way alignment of the ABCA molecule (Fig. 23a), the repeating unit of microdomains is simply of -ABC- type and the final morphology will be non-centrosymmetric. If, on the other hand, a two-way arrangement is achieved, the repeating unit exhibits an ABCACB pattern leading to a centrosymmetric structure (Fig. 23b).

A comparison of the calculated relative peak heights using the very simplified electron density profiles of Figs. 23b and 23d with the SAXS pattern of the sample, Fig. 24, unequivocally indicates the presence of the non-centrosysmmetric lamellae structure depicted in Fig. 23a. In agreement with SCMF calculations [90], the centrosymmetric two-way arrangement (Fig. 23c) of A-B-C-A-C-B pattern was not observed.

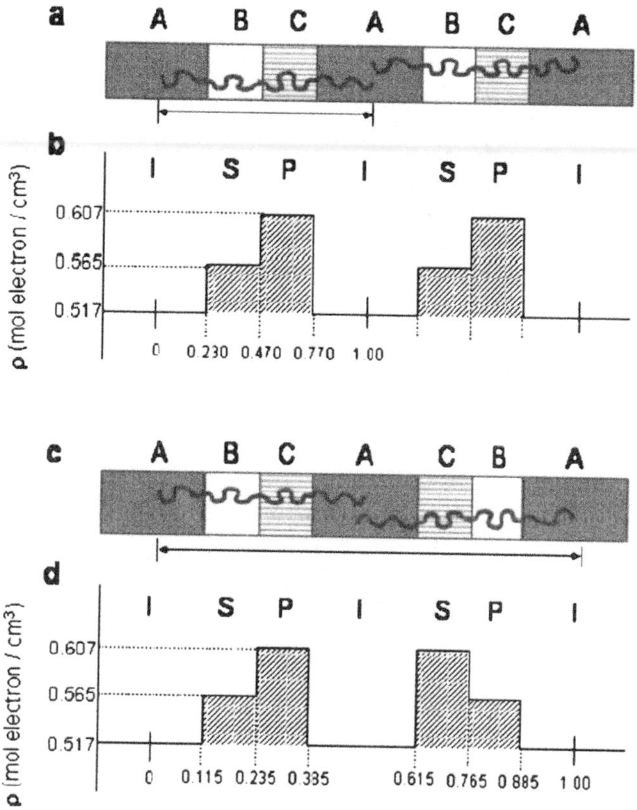

Fig. 23 Schematic comparison of non-centrosymmetric and centrosymmetric lamellar microdomain order. **a** One-way aligned pattern of ABCA molecules in microdomain structure. Repeating unit should be A-*b*-B-*b*-C, corresponding to PI-*b*-PS-*b*-P2VP. Darkness of each microdomain represents simplified TEM contrast: OsO_4 heavily stains PI phase, moderately stains P2VP phase, and very lightly stains PS phase. **b** Electron density profile corresponding to (**a**). Crystallographic unit cell is chosen as (PI/2)-*b*-PS-*b*-P2VP-*b*-(PI/2); in consequence several Bragg's peaks stem from this 1D array. This profile includes two-step stairs behaving as isolated asymmetric microdomains in matrix. **c** Two-way aligned manner of ABCA molecules, where repeating unit is A-*b*-B-*b*-C-*b*-A-*b*-C-*b*-B, i.e. PI-*b*-PS-*b*-P2VP-*b*-PI-*b*-P2VP-*b*-PS. **d** Electron density profile corresponding to (**c**). Crystallographic unit cell chosen here is -PI/2-*b*-PS-*b*-P2VP-*b*-PI-*b*-P2VP-*b*-PS-*b*-PI/2-. From [89]. Copyright 2003 American Chemical Society

2.2.14
Tetrablock Quarter Polymer

As a first member of multicomponent multiblock copolymers with more than three different blocks, a PS-*b*-PI-*b*-PDMS-*b*-P2VP copolymer was synthesized by Takahashi et al. [91]. As the constituents have different contrasts, the

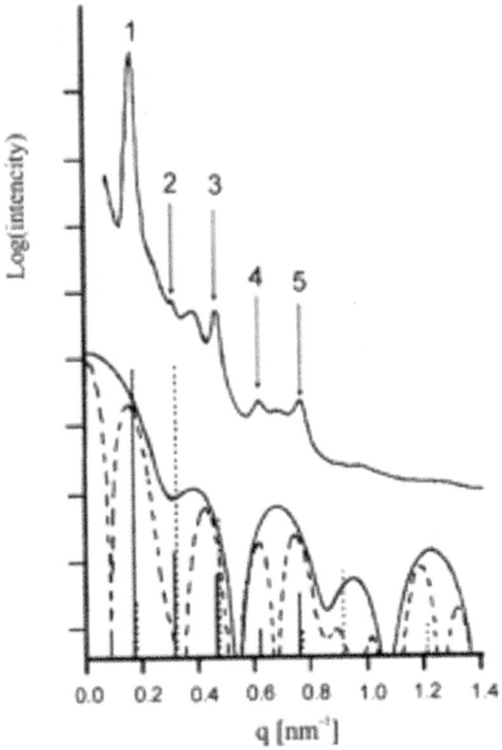

Fig. 24 SAXS pattern of PI-*b*-PS-*b*-P2VP-*b*-PI tetrablock terpolymer. Sample geometry is edge view with incident X-rays parallel to film surface. *Vertical thick bars* at lower section denote calculated diffracted intensities based on non-centrosymmetric periodic structure shown in Fig. 23b; *solid curve*: "domain" scattering calculated using profile of two-step stairs. *Vertical dotted bars*: calculated intensities based on centrosymmetric structure shown in (d); *dashed curve*: "domain" scattering for symmetric profile in same figure. From [89]. Copyright 2003 American Chemical Society

microphase-separated structure could even be observed without staining by TEM. The micrographs (Fig. 25) suggest a morphology in which the P2VP cylinders of a diameter of 60 nm assemble in a *hex* array. On the periphery of the cylinders the PDMS forms a layer-like domain exhibiting hexagonal shape. This cylindrical shell itself is covered by a PI domain which is embedded in the PS matrix with a 70-nm distance between the cylinder axes, as sketched in Fig. 26.

Recently, Takano et al. [92] published studies on the morphological behaviour of a polyisoprene-*b*-polystyrene-*b*-poly(2-vinylpyridine)-*b*-poly(4-trimethylsilylstyrene) (PI-*b*-PS-*b*-P2VP-*b*-PTMSS) tetrablock quarter polymer with unequal weight fractions ($\phi_{PI} = 0.25$, $\phi_{PS} = 0.23$, $\phi_{P2VP} = 0.32$ and $\phi_{PTMSS} = 0.20$). Here, the PI and PTMSS end blocks are almost compatible with each other, ($\chi_{PI\text{-}PTMSS} = 3.33 \times 10^{-3}$ at 418 K) offering the system two

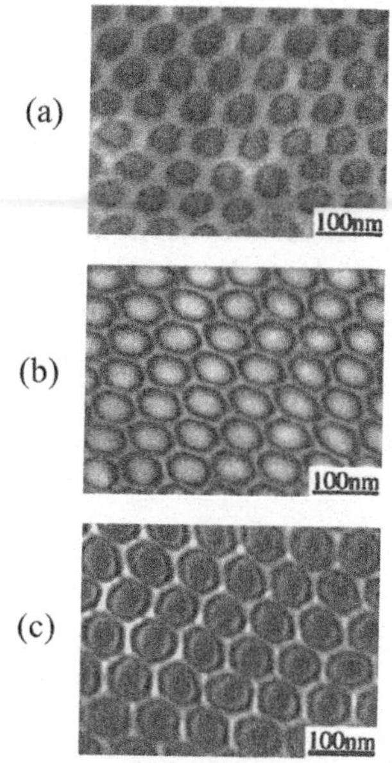

Fig. 25 Electron micrograph of annealed PS-*b*-PI-*b*-PDMS-*b*-P2VP **a** without staining, **b** stained with OsO_4 and **c** stained first with OsO_4 and subsequently with CH_3I. From [91]. Copyright 2002 American Chemical Society

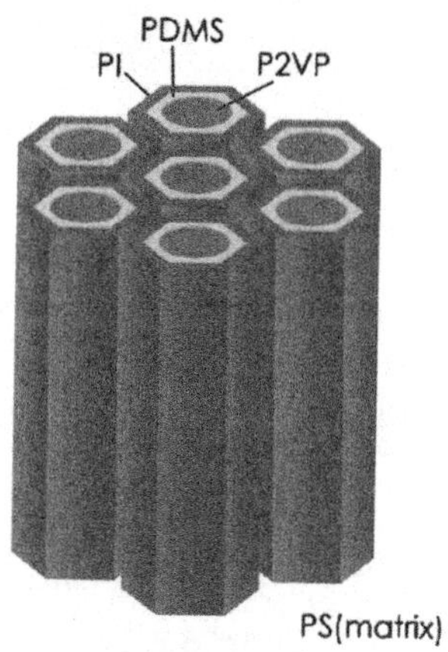

Fig. 26 Schematic illustration of model for hexagonal triple coaxial cylinder structure observed in PS-*b*-PI-*b*-PDMS-*b*-P2VP. From [91]. Copyright 2002 American Chemical Society

choices to assemble in a periodic structure. If the PI and PTMSS blocks belonging to different polymer chains are miscible under the conditions of sample preparation, the polymer aligns in a head-to-tail manner forming a non-centrosymmetric structure. It exhibits -(PI/PTMSS)-*b*-PS-*b*-P2VP- repeat units and assembles in a three-phase morphology, which is sketched in Fig. 27a. If, on the other hand, the PI and PTMSS block are immiscible, the system aligns in a head-to-head manner. This results in a centrosymmetric four-phase pattern with PI-*b*-PS-*b*-P2VP-*b*-PTMSS/PTMSS-*b*-P2VP-*b*-PS-*b*-PI repeat units as sketched in Fig. 27b. Results of SAXS and TEM clearly indicate the presence of the latter centrosymmetric arrangement.

However, recent theoretical studies [90] indicate that systems exhibiting attractive interactions of the A and D blocks generally should favour an -ABCDABCD- sequence leading to a non-centrosymmetric lamellar phase. As the theoretical phase diagram (Fig. 28) predicts that the annealing process goes through a mixed centrosymmetric phase (MCS BC) before the fully separated centrosymetric one (CS) is reached, the prediction may not be at

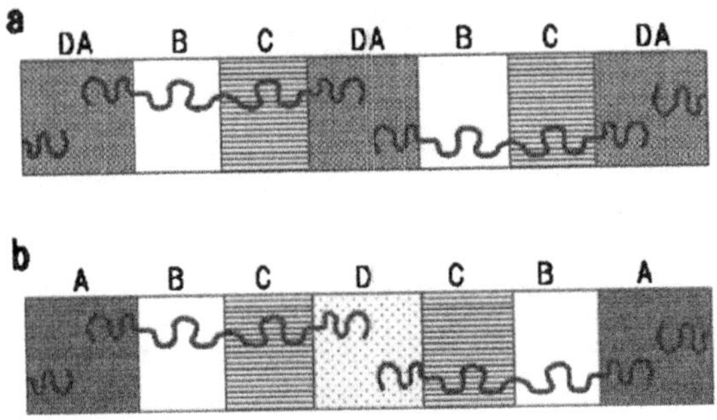

Fig. 27 Schematic drawings of two possible alignments of ABCD tetrablock quaterpolymer in lamellar microdomains: **a** ABCD and **b** ABCDCB. From [92]. Copyright 2003 American Chemical Society

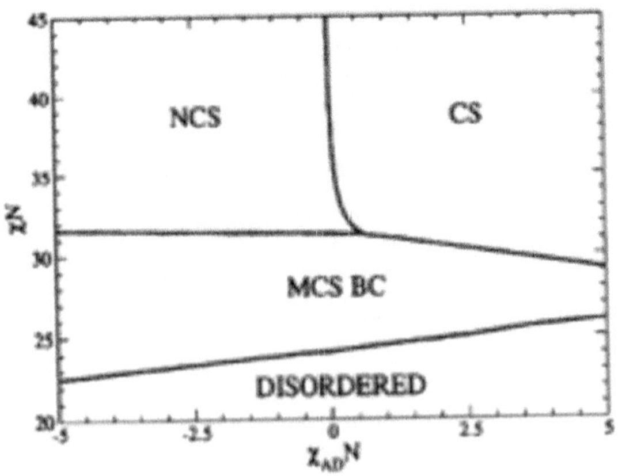

Fig. 28 Mean-field phase diagram for ABCD tetrablock quaterpolymer melts with $\phi = 1/4$. Flory–Huggins parameters are $\chi_{AB} = \chi$ except χ_{AD}. Phases abbreviations MCS BC: mixed centrosymmetric; NCS: non-centrosymmetric; CS: centrosymmetric. From [90]. Copyright 2004 American Chemical Society

variance with the experimental results. In addition, given the low value of $\chi_{PI\text{-}PTMSS}$, the system may also exhibit non-centrosymmetric lamellar structure as indicated in the centre region of Fig. 28.

3
Other Chain Topologies

3.1
Block Copolymers of Linear and Non-Linear Blocks

The morphology of a sixth-generation (G6) poly(benzyl ether) dendron covalently bonded to linear polystyrene (PS) at the dendron focal point has been evaluated by Mackay et al. [93] and is depicted in Fig. 29. Although the morphologies are progressing in an analogous fashion as in linear diblocks (cf. Fig. 1), they are distinct from those at equivalent volume fractions: Dendron cylinders were found when the dendritic volume fraction was 0.16, a value not outside but quite low compared to the minimum of 0.14 found for linear-linear systems in the SSL. The lamellar morphology assembles when the dendron volume fraction is 0.29, which is significantly lower than the minimum volume fraction of 0.36 at which lamellae are stable in linear-linear systems. Increasing the dendron volume fraction above 0.67 results in a microphase-separated material without long-range order [94].

Significant dendron shape distortion is required to accommodate the cylinder morphology in which the cylinder's diameter exceeds the diameter of the dendrons only by a factor of two. To allow formation of this geometry six single dendritic macromolecules have to assemble into a columnar

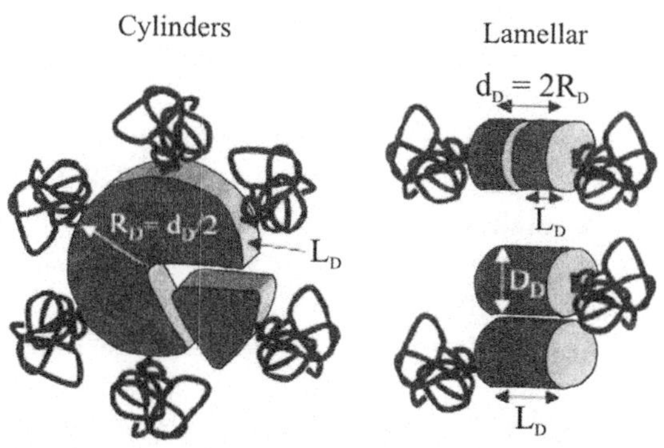

Fig. 29 Cartoon demonstrating arrangement of dendrons in hexagonally close-packed cylinder morphology (*left*) or in lamellar morphology (*right*). From [93]. Copyright 2002 American Chemical Society

morphology per cylindrical repeating unit. The lamellar morphology requires only slight chain stretches resulting in minor shape changes. In both morphologies the PS chains are slightly distorted. The results clearly indicate the key influence of molecular architecture on morphology.

3.2 Cyclic Block Copolymers

Linking two chain ends of a linear-block copolymer to each other yields a cyclic AB-block copolymer. The unique closed contour shape of the latter leads to distinctively different properties in solution and in melt compared to their linear analogues. Since the ring-shaped copolymers have no chain ends, their conformation in the microphase-separated state should be topologically different. They can only adopt a double-loop conformation. Both polymer blocks form loops in their respective domains, and the two junction points between the unlike blocks are located at the same interface. To compensate for the energetic penalty, the interface of the cyclic copolymer tends to curve away from the connected ends. The linear analogues of ABA type are able to form either single loops or bridged conformations. They may have their junction points confined at the same interface (loops) or at two different ones (bridges). This results in a larger domain spacing for linear-block copolymers, e.g. the spacing ratio, ρ, of a cyclic to its linear analogue diblock copolymer is calculated to be in a range from 0.6 to 0.7 [95, 96].

Lecommandoux et al. [96] investigated the effect of cyclization of a linear PS-*b*-PI observing a C morphology for the linear diblock and a liquid-like micellar phase for the cyclic one. The domain spacings, which were geometry-corrected to account for the different morphologies, were calculated from SAXS results. In agreement with theory [95], the cyclic morphology exhibited a smaller spacing with a typical ratio of $\rho = 0.707$ (Table 3). Furthermore, MFT was employed to obtain the ODT and the critical $(\chi N)_{ODT}$. The values obtained for cyclic copolymers were lower than those for the linear ones, though, which is in disagreement with theory.

The comparison of a linear PS-*b*-PI-*b*-PS triblock with its linear analogue was performed by Takano et al. [97]. As in the diblock case the cyclic copolymer exhibits a smaller domain spacing; however, the effect is not so pronounced (Table 3). This finding may be explained by the tendency of the ABA triblock to exhibit a higher curvature compared to an AB diblock, which in consequence reduces the differences between a cyclic diblock and its linear ABA counterpart.

For the PBut-*b*-PEO system the effect of cyclization has recently been reviewed by Ryan et al. (cf. Fig. 11) [61]. In analogy to PS-*b*-PI-*b*-PS, the stability of the disordered phase exceeds that of their linear counterparts of the same contour length [98]. Unlike in the former system, the interaction pa-

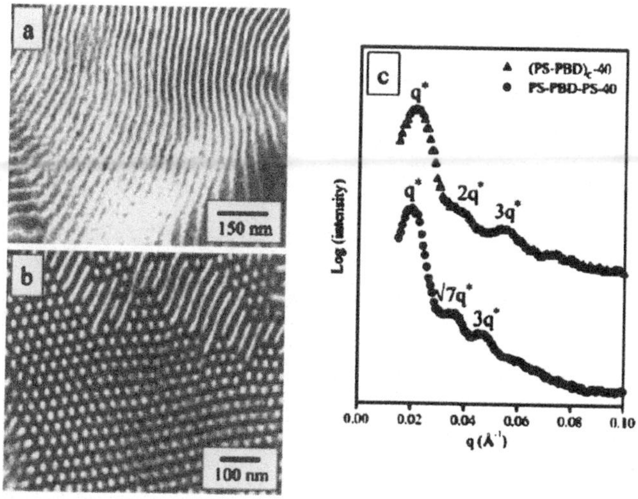

Fig. 30 Comparison of cyclic PS-*b*-PB and their linear PS-*b*-PB-*b*-PS analogue. Molecular weights were approx. 50 kg/mol; $\phi_{PS} = 0.4$. **a** TEM image cyclic PS-*b*-PB. **b** TEM image linear PS-*b*-PB-*b*-PS. **c** SAXS data, ▲: cyclic PS-*b*-PB; ●: linear PS-*b*-PB-*b*-PS. From [100]. Copyright 2003 American Chemical Society

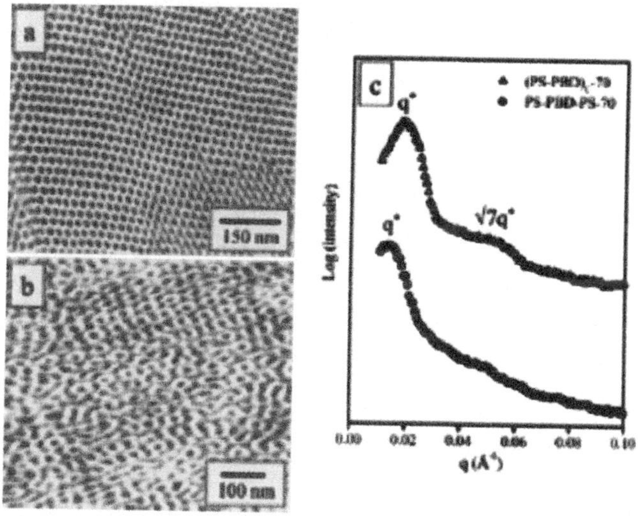

Fig. 31 Comparison of cyclic PS-*b*-PB and their linear PS-*b*-PB-*b*-PS analogue. $\phi_{PS} = 0.7$. **a** TEM image cyclic PS-*b*-PB. **b** TEM image linear PS-*b*-PB-*b*-PS. **c** SAXS data, ▲: cyclic PS-*b*-PB; ●: linear PS-*b*-PB-*b*-PS. From [100]. Copyright 2003 American Chemical Society

rameters of cyclic PBut-*b*-PEO are similar to the predictions of SCFT [95] and Monte Carlo simulations [99]. Again, linear analogues exhibit larger domain spacings compared to cyclic diblock copolymers.

Table 3 Comparison of the majority component's volume fraction, f, and the cyclic to linear spacing ratio, ρ, for different linear and cyclic block copolymers

Polymer	Morphology		Majority component	f	ρ	Ref.
	Cyclic	Linear				
PS-*b*-PI-*b*-PS	random *S*	*C*	PS	0.78	0.707	[96]
PS-*b*-PI-*b*-PS	*S*	*S*	PS	0.86	0.936	[97]
	C	*C*	PS	0.70	0.882	
	L	*L*	PS	0.51	0.867	
PBut-*b*-PEO	*L*	*L*	PEO	0.52 [a]	0.706	[61]
	L	*L*	PEO	0.52 [b]	0.687	
PBut-*b*-PEO-*b*-PEO	*L*	*L*	PEO	0.52 [a]	0.913	
	L	*L*	PEO	0.52 [b]	0.928	
PS-*b*-PB-*b*-PS	random *S*	random *S*	PB	0.89	0.889	[100]
	C	*C*	PB	0.76	0.844	
	L	*L*	PS	0.51	0.864	

[a]Overall number average molecular weight $M_n = 7100\ \mathrm{g\,mol^{-1}}$ [b]Overall number average molecular weight $M_n = 8100\ \mathrm{g\,mol^{-1}}$

Morphological differences between the cyclic PS-*b*-PB and their linear PS-*b*-PB-*b*-PS triblock analogues are highlighted in Figs. 30 and 31. Despite having an almost similar molecular weight and composition, the polymers assemble in a different structure, with the cyclic ones exhibiting the higher interfacial curvature [100]. For example, instead of PS cylinders for the linear-block copolymer, an *L* phase was found for the cyclic one (Fig. 30). The PB cylinders of a linear triblock copolymer became the minority domains of a *G* phase in the cyclic copolymer (Fig. 31). The change of long period or the change of morphology is attributed to an increased tendency of PS segments to orient parallel to the interface, when being end-linked together. In accordance with previous results, the linear triblock always shows larger domain spacing (Table 3).

The results for different domain spacings are summarized in Table 3. It is obvious that cyclic block copolymers exhibit smaller domain spacings compared to their linear analogues, which is due to an entropically less favourable situation and an increased interfacial curvature.

3.3 Star-Block Copolymers

The term "star-block copolymer" is used for a star architecture in which each arm is a diblock. The influence of chain topology on mechanical and morphological properties was investigated for copolymers composed of PS and PB with a constant styrene content of $\phi_{PS} = 0.74$ by Michler's group (Fig. 32) [101, 102]. While hexagonally packed cylinders of PB in a PS matrix were observed in a symmetric PS-*b*-PB-*b*-PS triblock copolymer, an *L* phase

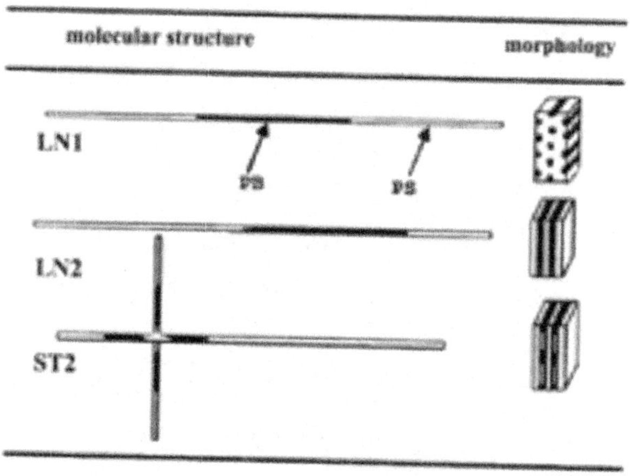

Fig. 32 Schematic representation of molecular structure and morphology observed in PS-*b*-PB-*b*-PS linear and star-block copolymers. *Oblique lines* between blocks for LN2 and ST2 indicate tapered transition of dissimilar blocks. From [102]. Copyright 2003 Wiley

was obtained for an asymmetric triblock copolymer of the same overall composition. A new so-called "two-component, three-phase" morphology, where additional PS domains are located within the PB lamellae, was observed in case of the tapered asymmetric star-block copolymer (Fig. 33). The symmetric PS-*b*-PB-*b*-PS triblock copolymer assembles in a *C* phase, though, and its asymmetric counterpart of the same composition exhibits a lamellar phase. In consequence, the unusual morphology of the star block may be attributed to a broadened interface arising from increased miscibility of both blocks driving the system toward weak segregation.

Blending the tapered star-block copolymer with PS further alters the morphology [103]. If the molecular weight of the blended homopolymer PS is much smaller than that of the longest PS star block but slightly higher than that of the smaller blocks, the homopolymer penetrates into the PS domains where it is solubilized. As a result, microphase-separated structures are formed. In analogy to the situation discussed in literature [104–107], this leads to an expansion of the PS domain size both laterally and normal to the interface and in consequence to an increase of the PS lamellae's thickness. As the molecular weight of the PS homopolymer approaches the molecular weight of the star's higher-molecular-weight blocks, the mixing entropy of the homopolymer chains is decreased. These chains become less successful to wet the copolymer brush effectively. Hence, the PS homopolymer and PS block

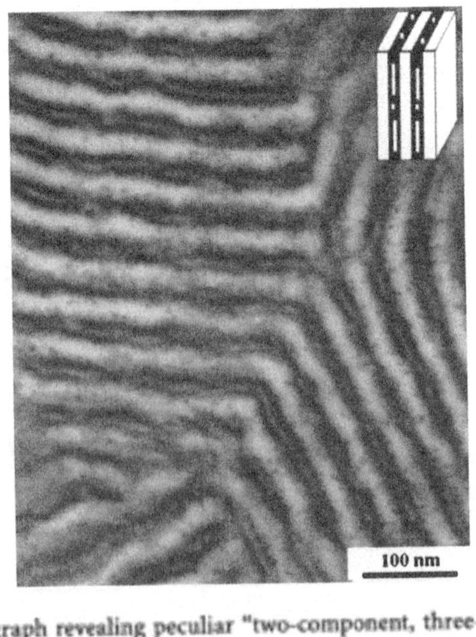

Fig. 33 TEM micrograph revealing peculiar "two-component, three-phase" morphology; OsO_4 staining. Note cylinder-like PS domains inside PB phase. Morphology is schematically illustrated in insert. From [102]. Copyright 2003 Wiley

become less miscible. As a result the copolymer's domain size shrinks. The PS homopolymer chains tend to be segregated in the middle of the PS domains or even completely expelled from the microdomains to form macrophase-separated PS particles. If the molecular weight of blended PS exceeds that of the longest PS block of the star, the PS homopolymers tend to phase-separate forming PS particles in block copolymer matrix.

The effect of the concentration of general-purpose PS with a broad molecular weight distribution on lamellar dimensions was shown in a subsequent publication [108]. The lamellar thicknesses as well as the lamellar spacings increase with increasing PS content. The major part of the added PS may be accommodated by the PS blocks of the star-block copolymer. For a PS concentration exceeding 60 wt %, the PS lamellae fuse together, and PS practically forms the matrix. Blending the star with 20 wt % of symmetric linear triblock copolymer consisting of two outer PS blocks held together by a random copolymer of PS and PB results in a worm-like morphology lacking long range order. With the destruction of the *L* structure, the small PS domains initially embedded in PB lamellae of the star block become less pronounced in the blends. Since a fraction of stars formed by coupling asymmetric PS-*b*-PB-*b*-PS molecules are butadiene rich and contain PS arms in the same range as the linear blend components, a common domain structure is formed. Partly, the linear and the star-block molecules are incompatible and phase-separate resulting in domains of different sizes and shapes. These domains have a microphase-separated morphology that resembles the structure of neither of the blend partners [109]. Unlike the classical rubber-modified or particle-filled thermoplastics, neither debonding at the particle/matrix interface nor the particle cavitation was observed in these nanostructured blends when being subjected to mechanical stress [110].

3.4
Miktoarm-Star Copolymers

In miktoarm-star copolymers (or "heteroarm star copolymers"), the unlike blocks are connected at one junction point, as shown in Fig. 34. As in linear-block copolymers, exactly two blocks are linked. Hence, it seems tempting to approximate those copolymers as a set of diblocks with the free ends of one block component joined together in a cluster [111].

However, the central junction of miktoarm-star copolymers exhibits more crowding compared to linear diblocks. The penalty for blocks bending at the interface will be higher (Fig. 34a,c), resulting in a shift of the phase boundaries and different spacings compared to their linear analogues of similar composition.

Recently, Grason and Kamien calculated the phase diagrams in the weak and strong segregation limit for AB_n miktoarm-star copolymers using both

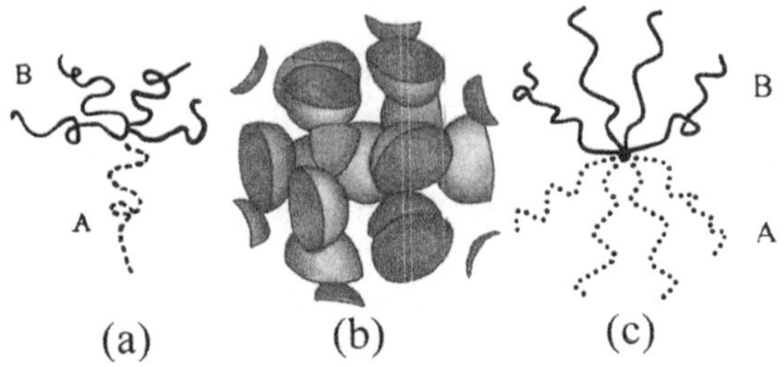

Fig. 34 Illustration of A_nB_n *miktoarm* star-block copolymer. **a** AB_4 diblock copolymer with A arm depicted by *dashed line* and B arms depicted by *solid lines*. **b** AB interfaces for A15 phase shown in $Pm\bar{3}n$ unit cell (extracted from SCFT results for $n = 5$ at $\chi N = 40$ and $\phi = 0.349$ [along *Hex*-A15 phase boundary] *vide infra*). From [112]. Copyright 2004 American Chemical Society. **c** A_nB_n *miktoarm*-star block copolymer with $n = 4$. From [121]. Copyright 2003 American Chemical Society

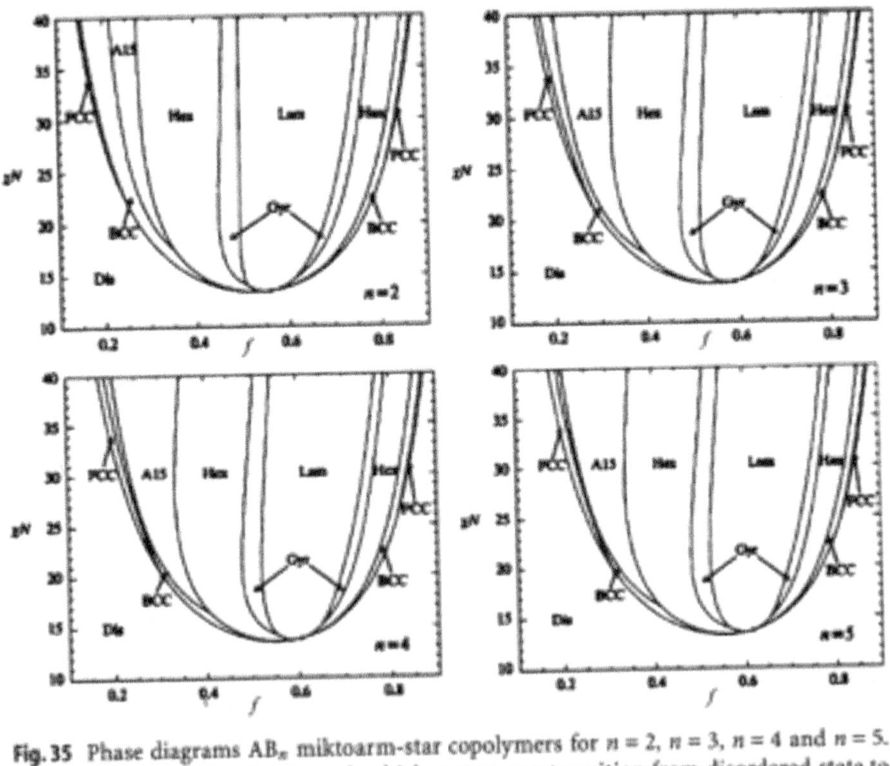

Fig. 35 Phase diagrams AB_n miktoarm-star copolymers for $n = 2$, $n = 3$, $n = 4$ and $n = 5$. •: mean-field critical point through which system can transition from disordered state to *Lam* phase via continuous, second-order phase transition. All other phase transitions are first-order. From [112]. Copyright 2004 American Chemical Society

a kinked-path approach and SCFT calculations [112]. The calculated phase diagrams for $1 \leq n \leq 4$ are depicted in Fig. 35.

For $\chi N \leq 40$, irrespective of the number of blocks, n, their results reveal non-symmetric phase diagrams about $\phi = 0.5$ (Fig. 35). Compared to the symmetric diagrams of their linear analogues boundaries are shifted to the B-rich region, indicating that phases with the n B-blocks on the outside of curved interfaces are favoured. This effect is driven by the additional asymmetry introduced by the molecular architecture and enhances with increasing n. A stable A15 phase of spherical micelles (Fig. 34b and Table 1) is observed. This morphology is unknown for classical linear diblocks (Fig. 1) and evidences a highly distorted AB interface.

In the strong segregation phase boundaries were calculated for $\chi N \approx 100$ and compared to the experimental results for PI-*arm*-PS stars. The results are shown in Fig. 36.

The computed results in Fig. 36 capture the observed phase behaviour of miktoarm-star copolymer melts. The effect of the molecular asymmetry seems to saturate for $n > 3$, and phase boundaries do not change significantly as a function of ϕ or higher asymmetry.

Investigations on a series of miktoarm-star AB_n copolymers of PS and poly(2-methyl-1,3-pentadiene), $PS(P2MP)_3$, indicate a different morphological behaviour [113] (Fig. 37), as predicted by Milner's theory [111] (Fig. 38). The discrepancies near the boundaries between different morphologies when compared with corresponding PS/PI systems on the one hand and to theoret-

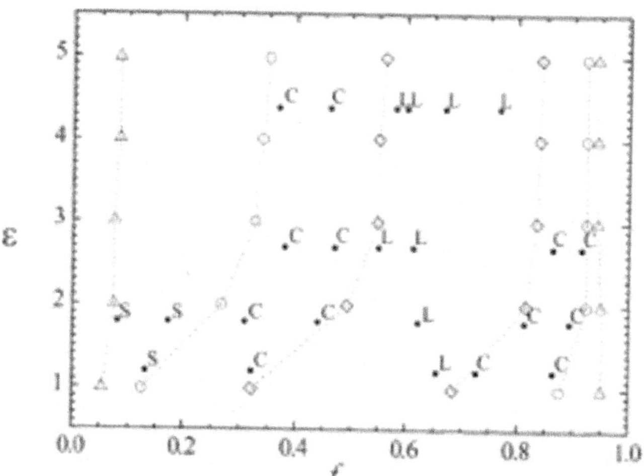

Fig. 36 SCFT results for AB_n miktoarm stars at strong segregation limit $\chi N = 100$. Phase transitions (△) Dis ↔ *bcc*; (○) *bcc* ↔ *Hex*; (◇) *Hex* ↔ *Lam*. All boundaries are computed at $\chi N = 100$ with exception of low-ϕ *bcc* ↔ *Hex* and *Hex* ↔ *Lam* ones for $n = 3$, 4 and 5. For $n = 3$ these were computed at $\chi N = 80$, and for $n = 4$ and 5 these boundaries are computed at $\chi N = 60$. •: Equilibrium results from experiments on PI-*arm*-PS melts [219]. From [112]. Copyright 2004 American Chemical Society

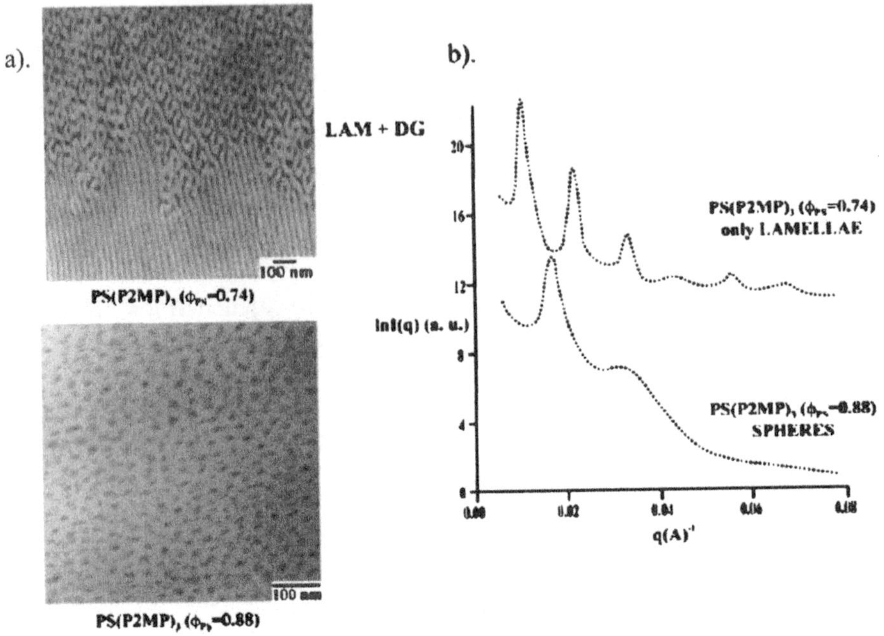

Fig. 37 Morphology of 4-miktoarm-star copolymers of $PS(P2MP)_3$ type. **a** TEM images. **b** SAXS data. From [113]. Copyright 2003 American Chemical Society

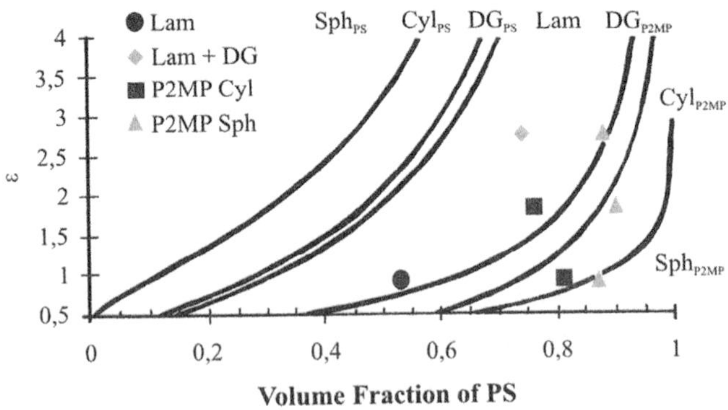

Fig. 38 Milner's diagram for morphologies of A-*b*-B copolymers as reported in literature [111]. Observed morphologies of linear and miktoarm-star copolymers of PS-*b*-P2MP system are represented. Lamellar structure, ■: cylinders of P2MP in PS matrix; ▲: spheres of P2MP in PS matrix, ♦: biphasic structure of lamellar/double gyroid microdomains. From [113]. Copyright 2003 American Chemical Society

ical predictions on the other are likely caused by differences in the characteristic ratios [114] and glass transition temperatures between P2MP and PI.

In the same study the coexistence of two morphologies (*L* and *G*) exhibited by a four-miktoarm-star copolymer was reported. This biphasic structure consists of very well ordered 1D lamellae and 3D substructures (core-shell gyroid) analogous to the results found in a three-miktoarm-star terpolymer of PS-*arm*-PB-*arm*-P2VP [115, 116] (Figs. 39 and 40).

Since a layered structure requires a certain segmental mixing in the neighbourhood of the junction point, these lamellar and core-shell gyroid morphologies are not expected for a three-miktoarm-star terpolymer composed of three incompatible blocks. For other compositions of this three-miktoarm-star terpolymer, tetragonal and hexagonal morphologies were found in which the junction point is located along an interfacial line rather than being located on an interfacial plane [5, 117].

The differences in chain conformations and interfacial structures between a linear PS-*b*-P2VP-*b*-PS triblock and a PS-*arm*-P2VP-*arm*-P2VP star copolymer was investigated by Torikai et al. [118]. Although their SANS data did not permit the distinction between "looped" and "bridged" conformation, they anticipated that the middle block of the linear copolymer would preferentially exhibit the former conformation for high f_{PS} and the latter one for low ϕ_{PS}. The interfacial thickness of the linear triblock is approximately constant irrespective of ϕ_{PS} and comparable with that of the PS-*b*-P2VP diblock copolymer. This implies that the difference in the conformation of the middle block chain does not have much influence on the interfacial structure. The PS chains of the graft copolymer have slightly larger dimensions along the direction parallel to the lamellar interface than the corresponding chains of the diblock. The chains were found to be less deformed for compensating asymmetric chain distributions at the interface.

The influence of the architecture on the phase behaviour of symmetric miktoarm stars A_nB_n (Fig. 34c) was investigated by Grayer et al. [119]. Symmetric miktoarm-star copolymers PS-*arm*-P2VP having a mean func-

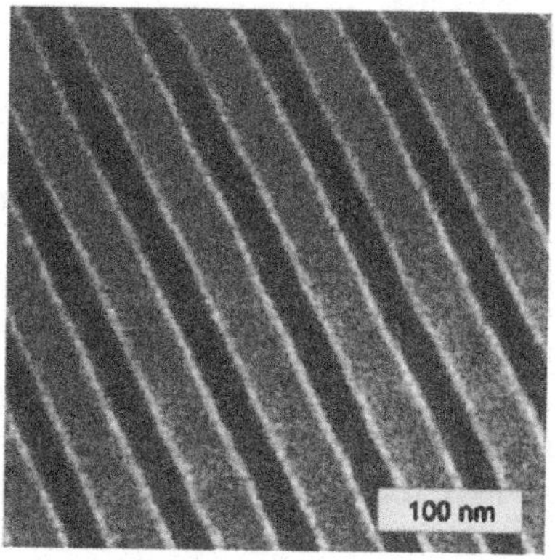

Fig. 39 TEM micrograph of lamellar region of PS-*arm*-PB-*arm*-P2VP miktoarm-star block copolymer stained with OsO_4 and CH_3I. From [115]. Copyright 2000 American Chemical Society

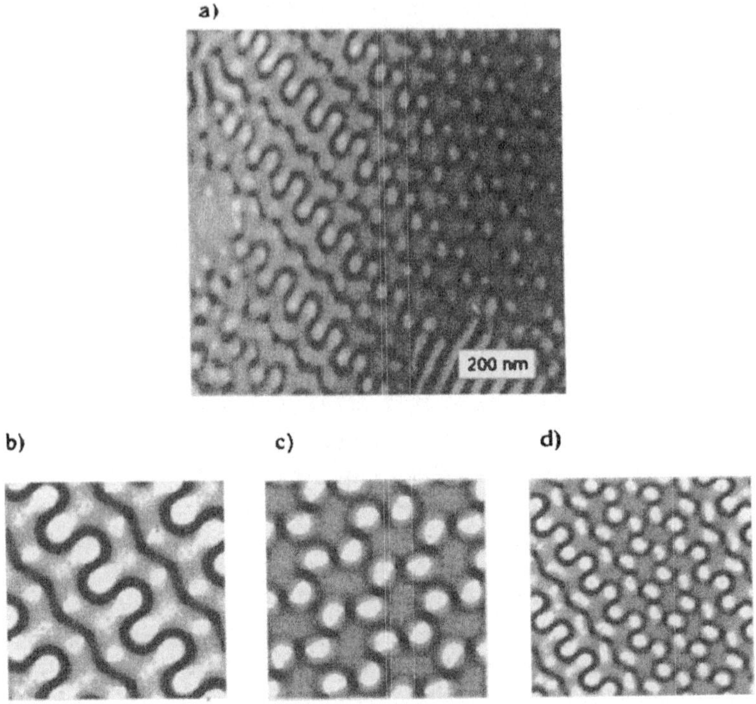

Fig. 40 **a** TEM micrograph of PS-*arm*-PB-*arm*-P2VP miktoarm-star block copolymer stained with OsO_4. **b** Simulation of [112] projection (*dark matrix*, translation 0, thickness 0.33). **c** Simulation of [112] projection (*dark matrix*, translation 0.45, thickness 0.33). **d** Simulation of "[1.08; 0.84; 2.04]" projection (*dark matrix*, translation 0.48, thickness 0.35). From [115]. Copyright 2000 American Chemical Society

tionality of 12 were employed. By keeping the length of the PS arms nearly constant at $M_w(PS) \approx 1.55 \times 10^4$ and varying the length of the P2VP arms, a composition range of $0.30 \leq \phi_{PS} \leq 0.75$ was covered. The phase space of the samples was compared to the theoretical phase diagram calculated by Matsen et al. [20] for diblock copolymers in the intermediate segregation regime (Fig. 41). Three of the five samples investigated by TEM lie in the region of stability of the *L* phase. One showed a structure made of small regions of *C*. Both findings agree well with theory. The sample with the lowest ϕ_{PS} showed no particular ordering, indicating a bicontinous structure or a non-equilibrium frozen state despite the fact that theory predicts a columnar array. The lamellar spacing of the miktoarm stars exceeds the spacing of their linear counterparts of 20–30%, indicating an additional stretching of the arms near the interface. Further analysis revealed that this difference is the most pronounced for stars of large functionality with small and intermediate lengths of the P2VP block ($0.8 \leq M_w(P2VP)/10^4 \leq 4.1$). For the stars with longer P2VP arms ($M_w(P2VP) \approx$

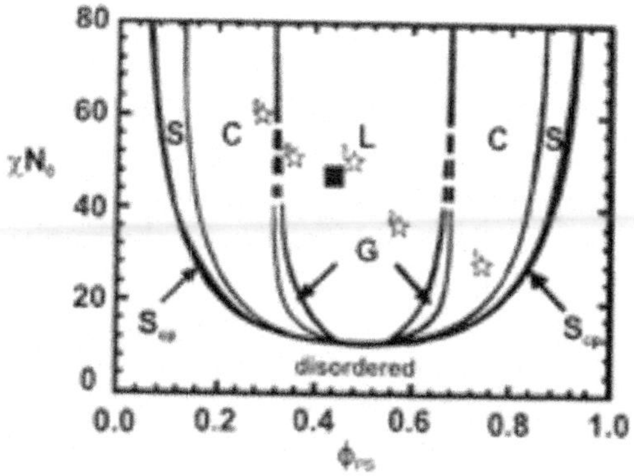

Fig. 41 Theoretical phase diagram calculated by Matsen et al. [20] for diblock copolymers in intermediate segregation regime. ☆: PS-*arm*-P2VP miktoarm polymers; ■: PS-*b*-P2VP linear polymer. From [119]. Copyright 2000 American Chemical Society

5.5×10^4), however, the difference with the diblock copolymers becomes smaller.

For symmetric PS-*b*-PI the differences between linear AB diblock copolymers and their miktoarm-star A_2B_2 analogues were studied by Buzza et al. [120]. Both the three linear and the corresponding star polymers of arm-molecular weights of 10, 20 and 60 kg/mol exhibited *L* morphology. The dependency between the spacing determined from the position of the first-order SAXS reflection peak, q^*, and the overall molecular weight, *N*, follows the scaling law $q^* \sim N^{-a}$ with $a \approx 0.7$ for both the linear and miktoarm star (Fig. 42). However, in the latter case the chains were consistently 5–10% more

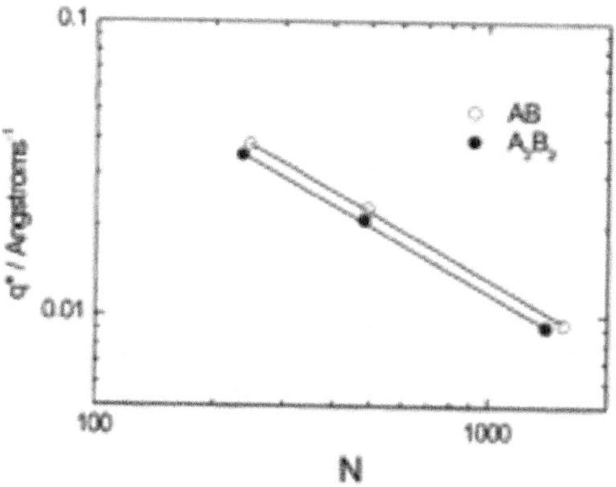

Fig. 42 Double-logarithmic dependence of first-order reflection peak, q^*, on total degree of polymerization, *N*. ○: AB diblocks; ●: A_2B_2 miktoarm stars. *Solid lines*: fitting results for power law scaling of $q^* \sim N^{-a}$, with $a = 0.73 \pm 0.04$ and 0.70 ± 0.04 for linear and stars respectively. From [120]. Copyright 2000 American Chemical Society

stretched as compared to their linear analogues, indicating stronger phase segregation. As the rheologically determined T_{ODT} of the miktoarm star exceeds the T_{ODT} of its linear counterpart by 40 °C (arm molecular weight 10 kg/mol), a stronger phase segregation due to the introduction of a branching point can be anticipated.

The dependence of the lamellar thickness and the number of arms ($n = 1$, 2, 4 and 16) for symmetric PS_n-*arm*-PI_n miktoarm stars shows an increase in the spacing with n (Fig. 43). This indicates an additional chain stretching induced by the spatial confinements close to the junction point. However, the exactness of the results may be influenced by non-separable impurities. As these contamination species are resistant to detection via standard SEC and other separation techniques, it can be reasoned that previous results reported in the literature might suffer from the same shortcomings [121].

In conclusion, it can be suggested that the lamellar spacing of miktoarm-star copolymers is controlled by two parameters: (i) the molecular weight of the corresponding AB diblock, which also controls the segregation strength, and (ii) the functionality of the central core.

In analogy to linear-block copolymers different cases can be distinguished when blending asymmetric miktoarm (PS-PI)$_n$-PS and H-shaped (PS-PI)$_3$-PS-(PI-PS)$_3$ copolymers with homopolymer PS [122]. If the latter's molecular weight is lower than the respective PS block, a transition from the L structure to hexagonally packed cylinders without observation of an intervening cubic morphology is observed in the case of the (PS-PI)$_n$-PS types. If the H-shaped (PS-PI)$_3$-PS-(PI-PS)$_3$ star-block copolymer is blended with 30% to

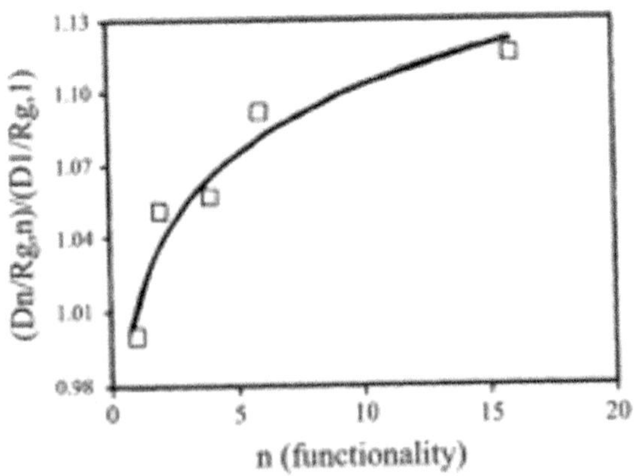

Fig. 43 Plot of normalized lamellar long periods, $D_n/R_{g,n}$, of (PS)$_n$-*arm*-(PI)$_n$ miktoarm-star copolymers ($n = 1$, 2, 4 and 16) divided by corresponding diblocks of same series, $D_1/R_{g,1}$, against respective star functionality, n. Normalization factor $R_{g,n}$ represents unperturbed radius of gyration of diblock consisting of one PS block, one PI block and average number of bonds linking these two arms through core. From [121]. Copyright 2003 American Chemical Society

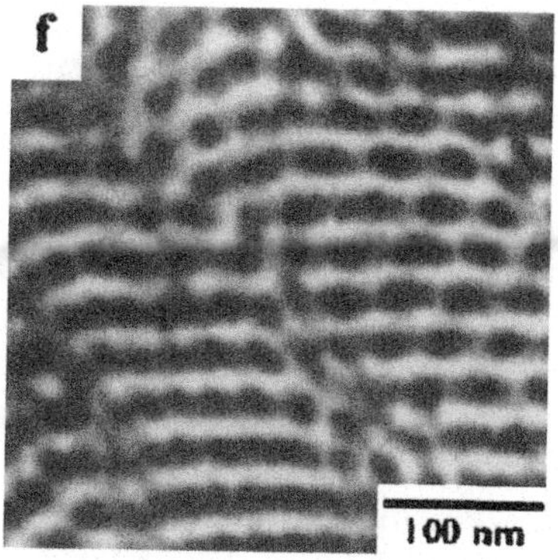

Fig. 44 TEM image for binary blend of H-shaped $(PS\text{-}PI)_3\text{-}PS\text{-}(PI\text{-}PS)_3$ with 35 wt % PS stained with OsO_4 in bright field mode. From [122]. Copyright 2002 Elsevier

35% of homopolymer, it self-assembles to a lamellar-*PL* structure, as depicted in Fig. 44.

The TEM (Fig. 44) and SAXS data clearly indicate that the structure is lamellar to the first order but modified by channels in the PI layers, leading to a *PL* pattern. However, Fredrickson's theoretical results indicate that in the SSL this structure is not the most favourable one [123]. In the WSL, similar *PLs* are to occur for equal weight fractions, even though lamellae are expected in that region [124]. A similar behaviour has been observed in the case of PI_2PS homopolymer blends for the *PL* morphology [15].

For H-shaped as well as the $(PS\text{-}PI)_n\text{-}PS$ case the addition of low molecular weight PI homopolymer did not change the lamellar morphology, even for highly asymmetric compositions. The effect of the different homopolymers was explained as follows. Addition of the PS increases the thickness of the PS layer while decreasing that of the PI one. The opposite trend occurs when swelling the system with PI. However, measurements of domain spacing showed an analogy to the behaviour of a $PS\text{-}PI_2$ graft block copolymer [15], where the PI homopolymer did not penetrate as well into the two-arm-per-molecule PI regime as did the PS homopolymer into the PS side of the interface. This asymmetry introduced by the asymmetric architecture leads to the observed splitting of the OOTs.

Despite the fact that Milner's theory was originally developed for miktoarm-star copolymers, it can also be adopted for more complex branched structures. This empirical concept termed "constituting-block copolymers" approximates the architecture of branched molecules to be composed of an array of A_2B and A_2B_2 miktoarms. This approach is capable of predicting the morphology of architectures as complex as centipedes or barbed wires, as shown in a very recent publication [125].

4
Crossing the Boundaries Between Different Phases

Phase transitions in block copolymers can be either order-order transitions (OOT) or order-disorder ones (ODT). The first type includes changes between the various ordered morphologies discussed in Fig. 1, whereas the latter describes the transition between the presence and absence of long-range interactions in the system. For a block copolymer of given molecular weight and composition, both transitions may be triggered by variation of the temperature and/or pressure. Blending offers another possibility to alter the system's phase space. Besides the temperature/pressure the variation of the different component's fraction offers an elegant (and most used) possibility to tailor the system's phase space.

4.1
Kinetics of Transition Processes

The viscoelastic effects on the morphology and dynamics of microphase separation of diblock copolymers was simulated by Huo et al. [126] based on Tanaka's viscoelastic model [127] in the presence and absence of additional thermal noise. Their results indicate that for $\phi = 0.5$ and an identical bulk modulus of both blocks, the area fraction of the A-rich phase remains constant during the microphase separation process. For each block randomly oriented lamellae are preferred.

In the case of bulk modulus differences between the two blocks the area fraction of the higher bulk modulus A-rich phase reaches a maximum during separation and then decreases to a steady-state value that is slightly different from ϕ. The morphology exhibits a thin network-like continuous phase of the A material with the lower bulk modulus B-rich droplets dispersed in the A-rich matrix. In symmetric systems the viscoelastic effect does not cause a variation in the computed morphologies. For $\phi_A = 0.4$ the situation is different, which is due to inherent composition asymmetries. Although the minority component A possesses a higher bulk modulus, it tends to form a thin lamellar morphology. From the pattern evolution the lower bulk modulus B-rich droplets begin to emerge first, and then grow in size, with the A-rich phase being compressed to form a network-like structure. As the microphase separation proceeds further, the B-rich phase eventually forms the continuous phase and the A-rich network is partially broken up. It is obvious that the minor component A has a greater tendency to form a lamellar morphology (Fig. 45a). On the other hand, if the minority component A possesses a lower bulk modulus, the domains of the A-rich phase first emerge from the B-rich matrix and then eventually evolve into isolated A-rich droplets (Fig. 45b). The morphological differences between the absence and the presence of bulk modulus differences emphasize the importance of accounting

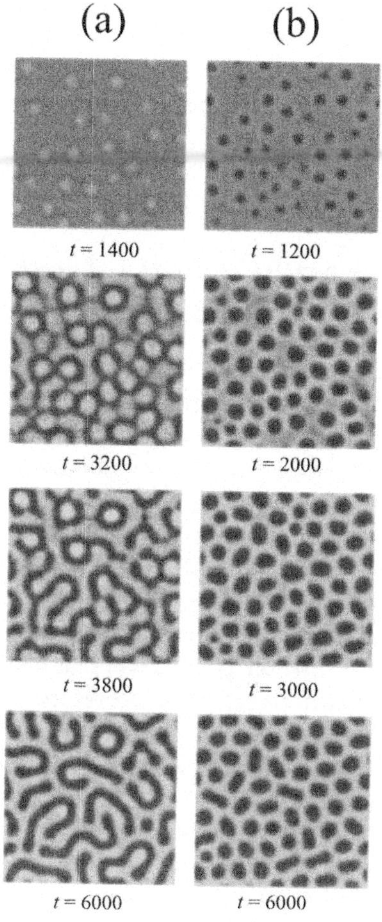

Fig. 45 Simulated pattern evolution during microphase separation for $\phi_A = 0.4$ with A having a 10 times higher (**a**) and lower (**b**) bulk modulus than B. *Black*: A-rich regions; *white*: B-rich regions. From [126]. Copyright 2003 American Chemical Society

for the viscoelastic properties. They may be responsible for the deviations between the experimentally measured and the mean-field theoretically predicted phase diagrams.

Small-angle X-ray studies indicate that long equilibration times after temperature jumps within the ordered state of sphere-forming PI-*b*-PDMS and PEP-*b*-PDMS copolymers are due to slow chain diffusion. The diffusion changes the number of domains (if *T*-jump occurs within one phase) or the symmetry of domains (if an OOT is involved). Measurements using tracers indicated that chain mobility was not rate-limiting, although variations in mobility did account for the differences between the two investigated polymers. Moreover, the required reduction in the number density of spheres on cooling was found to be rate limiting. The disappearance of spheres, either by "evaporation" or by "fusion", provided a large kinetic barrier. Lamellae, however, could adjust domain dimensions simply by local displacements of individual chains and for this reason occur rather rapidly [128].

4.2
Transitions Between Disordered and Ordered States

Usually the discussion of the ODT of highly asymmetric block copolymers in the strong segregation limit starts from a body-centred cubic (*bcc*) array of the minority phase. Phase transitions were calculated using SCFT accounting for both the translational entropy of the micelles in a disordered micelle regime and the intermicelle free energy [129]. Results indicate that the ODT occurs between ordered *bcc* spheres and disordered micelles.

Contrary to the more symmetric *L* or *C* systems which transform directly to the disordered state with thermally induced composition fluctuations at T_{ODT}, the highly asymmetric spherical ones undergo transformation to disordered micelles. These exhibit liquid-like short-range order disappearing at even higher temperatures, resulting in an amorphous melt. To emphasize the distinction, the transition from *bcc* to disordered micelles is termed the lattice disordering transition (LDT), whereas the disappearance of the micelles occurs at the demicellization temperature (DMT) [130–132]. This definition, however, leads to the conceptual difficulty of relating the DMT to the ODT, which is completely inconsistent with the requirement that the ODT is a true first-order thermodynamic phase transition taking place between an ordered and a disordered state. Independent of the definition, the temperature-dependent phase behaviour of an asymmetric PS-*b*-PI system was predicted, and reasonable agreement with the experimentally observed temperatures of ODT and LDT was found [129, 132].

Above the ODT the number density of micelles decreases steadily. Over some region, termed the critical micelle temperature (cmt), the rate of decrease changes, and the free energy of the disordered micelle regime (with very low micelle concentration) is only slightly lower than the Flory–Huggins free energy for a homogeneous melt. Hence, the cmt may be defined as the temperature which separates the disordered micelle regime (with a finite micelle concentration) from a disordered melt (with an exponentially small micelle fraction); however, it is not a true phase transition. At even higher temperatures the formation of micelles becomes thermodynamically less favourable, and their fraction decreases considerably. The micelle concentration will always be non-zero as even an infinitesimally small fraction of micelles allows the system to gain translational entropy. The exact definition of the transition process between ordered *bcc* and random micelles on one hand and randomly distributed micelles and the disordered melt on the other is still under discussion. Some authors argue that the existence of the disordered micelle regime represents an essential difference between sphere-forming and more symmetric block copolymer systems and therefore requires a special treatment.

The architecture as well as the preparation of the test specimen has a pronounced effect on the ODT. Choi et al. [133] found that T_{ODT} of the symmetric PS-*b*-PI-*b*-PS copolymer is approx. 20 °C higher than that of the corresponding PS-*b*-PI block copolymer, which is in agreement with the previous experimental results [134]. In the same study the preparation by compression molding or solvent casting was found to influence T_{ODT} and demicellazation transformation in the case of highly asymmetric PS-*b*-PI and PS-*b*-PI-*b*-PS respectively, while the ODT for lamellar or cylindrical block copolymers was less influenced [133]. Based on the researchers' findings they recommend a practice for determining the phase-transition temperature(s) of highly asymmetric block copolymers: (i) a slow solvent casting, instead of compression molding, should be used to prepare specimens, (ii) optimum annealing conditions in terms of temperature and the duration of annealing should be determined and (iii) sufficiently low angular frequencies (e.g. 0.01 rad/s) should be applied to a specimen when conducting isochronal dynamic temperature sweep experiments.

The ordering kinetics was investigated for PS-*b*-PI-*b*-PS triblock copolymers after *T*-quenches to the *S* or *C* phase [135]. After deep quenches, the system stays in the supercooled disordered-sphere phase for a certain incubation period. Thereafter the grains composed of *Hex*-cylinders nucleate in the disordered-sphere phase and grow at the expense of the disordered-sphere phase, i.e. the *C* phase is formed directly from the disordered spheres. Shallow quenches into the *C* phase proceed via a different pathway: The system stays in the supercooled disordered-sphere phase for a certain incubation period, too. Then the grains composed of *bcc* spheres appear in the disordered-sphere phase and grow at the expense of it. The ordered volume filling *bcc* spheres are metastable and will undergo OOT into the final *C* morphology when the complete sample space is filled with them.

4.3 Transitions Between Ordered States

The order-order transition (OOT) of a block copolymer is defined as the change of one type of ordered microdomain structure to another one. In the strong segregation limit it basically depends on the volume fraction of one of the constituent block chains, ϕ. In the WSL it is also influenced by χN and therefore may occur by changing the temperature. The thermoreversible OOT between *S* and *C* of a PS-*b*-PI block copolymer with a PS weight fraction of 0.20 was investigated by Kimishima et al. [136]. By using SAXS and polarized optical microscopy they observed cylindrical microdomains on a hexagonal lattice up to 114.7 °C, which changed reversibly to *bcc* spheres above 116.7 °C.

A straightforward transition process [137] from the *C* to *S* was proposed (Fig. 46): The cylinder breaks up into spheres with the original cylindrical axis parallel to the [111] direction of the *bcc* sphere lattice induced by thermodynamic instability of the cylindrical interface at $T > T_{OOT}$. The process may proceed via four subsequent steps: (i) undulation (as shown from part a to b), (ii) break-up of cylinders into ellipsoids (from part b to c), (iii) relaxation of domains from the ellipsoids into spheres (from part c to d), and (iv) relaxation in a junction distribution to attain a uniform distribution (from part d to e). These processes involve only local rearrangements of the block copolymer molecules rather than large length-scale rearrangements of block copolymer molecules or microdomains. The reverse transition involves deformation and elongation of the spheres into ellipsoids and coalescence of

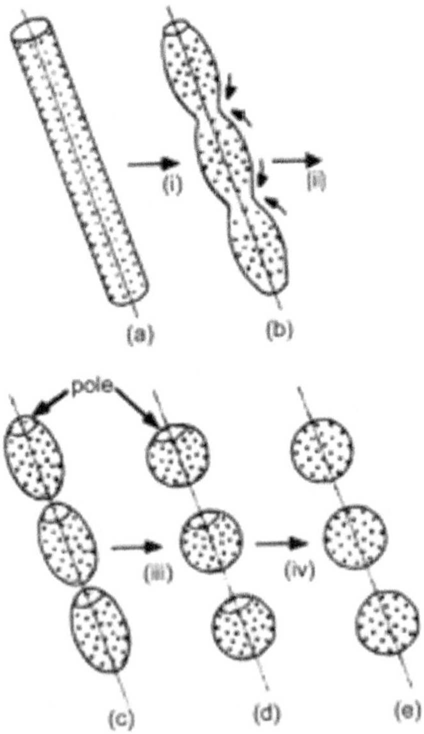

Fig. 46 Schematic diagram of elemental process during transition from *Hex* cylinder to *bcc* sphere; (i) undulation of interface (**a**, **b**), (ii) break-up of cylinders into ellipsoids (**b**, **c**), (iii) relaxation of domains from ellipsoids into spheres (**c**, **d**), and (iv) relaxation in junction distribution to attain uniform distribution (**d**, **e**). Pole where concentration of junction points is low may work as memory of grain conservation upon reverse transition from *bcc* sphere to *Hex* cylinder. *Small arrows* in part (**b**) indicate diffusion of chemical junctions along interface in process (ii). From [136]. Copyright 2000 American Chemical Society

the ellipsoids into a cylinder. These processes are driven by thermodynamic instability of spherical interfaces caused by increased segregation power at $T < T_{OOT}$.

The mean-field phase diagram in the WSL calculated by Matsen et al. [138] predicts a transition from *C* to the disordered state via the *bcc* and the *fcc* array with decreasing χN. This was not followed here. Transitions from the *C* to *S* (at 115.7 °C), to the lattice-disordered sphere – where the *bcc* lattice was distorted by thermal fluctuations – and finally to the disordered state (estimation > 180 °C but not attained in the study) were observed. It was reasoned to consider the lattice-disordered spheres as a fluctuation-induced lattice disordered phase. This enlarges the window for the disordered one and causes the *fcc* phase to disappear. Even if the latter should exist, its observation will be aggravated by its narrow temperature width of about 8 K and its slow formation due to the symmetry changes between *fcc* and *bcc* spheres.

Reversible *C* to *S* transitions were examined for PS-*b*-PI and PS-*b*-PI-*b*-PS block copolymers through several repeated heating/cooling cycles [139]. An unexpected orientational proliferation of the *C* phase and successive twinning of the *S* phase was observed during these cycles. In the case of the *C* to *S* transition, epitaxy dictates that one of the directions [111] of the *S* phase coincides with that of the cylinder axis of the original *C* phase. The other three [111] directions form the skeletons of a tetrahedron with the first one, with their azimuth orientation conforming to the hexagonal symmetry of the cylinders. Furthermore, it was both predicted theoretically [140] and demonstrated experimentally [141, 142] that the transition from *C* to *S* yields two *bcc* structures connected by twinning as opposed to a single *bcc*. Upon the reverse transition, each of the [111] directions of the *bcc* can become the cylinder axis in the *C* phase. Since there are four [111] equivalent directions in each of the *bcc* twins, with one of the four (along the initial cylinder orientation) shared by both, seven distinct *C* orientations are to be expected after one *C* → twinned *bcc* → *C* cycle (Fig. 47).

Starting from an initially aligned cylinder, *C*-0, seven cylinder orientations (including the initial one) are created after the first heating/cooling cycle and 25 cylinder orientations (including the seven present in *C*-1) are generated after the second cycle as indicated by 2D-SAXS, rheo-optics, and TEM. The non-zero birefringence of *C*-1 suggests a preference for the cylinder orientation in the initially shear-aligned direction, which indicates some memory of the initially aligned *C* state. This memory effect is shown to decrease with annealing time in the twinned *bcc* state. The authors speculate that differences in the memory effects may be responsible for the fact that previous studies did not observe these effects.

The pathway and kinetics of the *C* to *S* transition have been studied on shear-aligned cylinders of the commercial diblock copolymer of PS and poly(ethylene-*co*-butylene) (KRATON G 1657 Shell Chemical Company) [143, 144]. A complete dissolution of the cylindrical structure before the epitaxial

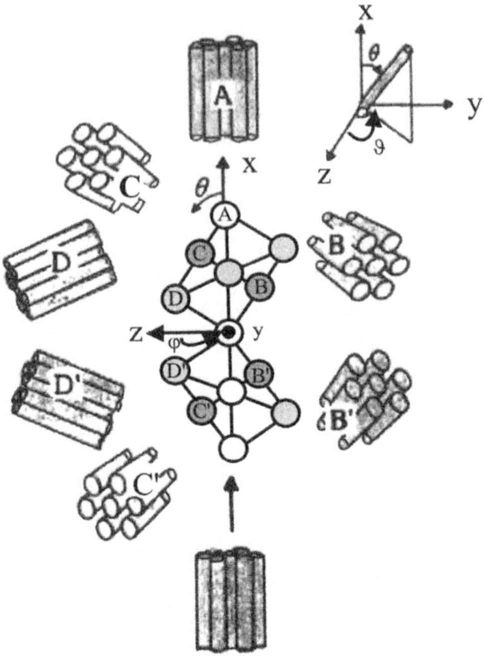

Fig. 47 Schematic of transition *C*-0 → twinned *bcc*-1 → *C*-1. From [139]. Copyright 2002 American Chemical Society

growth of the spherical microdomain structure was observed, which is qualitatively different from the pathway elucidated from previous studies [141, 142, 145–148]. The discrepancy could be caused by a mismatch in the interdomain spacing for the cylindrical and spherical microdomains, or a smearing of the transition due to the examination of a blend of a matched diblock and triblock copolymer.

The kinetics and mechanisms of the *C* → *G* transition in a concentrated solution of PS-*b*-PI in the PS-selective solvent di-*n*-butyl phthalate was studied [137, 149]. An epitaxially transformation of the shear-oriented *C* phase to *G*, as previously established in melts [13, 50, 150], was observed. For shallow quenches into *G*, the transition proceeds directly by a nucleation and growth process. For deeper quenches, a metastable intermediate structure appears, with scattering and rheological features consistent with the hexagonally perforated layer (*PL*) state. The *C* → *G* transition follows the same pathways, and at approximately the same rates, even when the initial *C* phase is not shear-oriented.

Phase transitions of the PS-*b*-PI system have been extensively studied. The morphological transition from the *L* phase to the *G* phase proceeds through nucleation and growth. The difference in the geometrical characteristics of these two phases induces considerable local distortion of both morphologies during the transition. The resulting strain raises the surface energy of the grains, leading to a suppression of the direct $L \rightarrow G$ transition in case of an asymmetric polymer ($\phi_{PI} = 0.3$). Instead, the *PL* phase tends to appear as an intermediate structure capable of forming low-energy grain boundaries during the $L \rightarrow G$ transition.

The overall arrangement of the *PL* phase retains a planar aspect reminiscent of the *L* phase, though the minority components are already arranged in a planar, three-fold coordinated structure. Hence, the *PL* phase combines characteristics of the *L* and *G* phase, which accounts for its intermediate role [12]. The *PL* structure is formed through the development of periodic distortions in the interface of the lamella which separates the blocks. The distortions grow in amplitude until channels extending through the minority component domains are formed. The $PL \rightarrow G$ transition can proceed with epitaxial relation via nucleation and growth because both *PL* and *G* morphologies are constructed from a nearly identical, three-fold-coordinated minority component [151].

However, a PS-*b*-PI/PI blend shows direct $L \rightarrow G$ transitions without appearance of the *PL* phase. The *L* microdomain is more favourable than the *PL* phase since the volume fraction of the PI block component and the symmetry of microdomains is increased by the addition of PI homopolymer. Hence, the *PL* phase may not be formed as an intermediate structure if relatively high molecular weight PI homopolymer is added. The latter is not able to effectively fill the corners of the Wigner–Seitz cells; in consequence packing frustration cannot be released and the PL phase is not favoured [152]. In contrast, the addition of low molecular weight PI homopolymer to the minor component of the *PL* phase reduces the packing frustration imposed on the block copolymers and stabilizes it [153]. Hence, transition from the *PL* to the *G* phase indicates an epitaxial relationship between the two structures, while the direct transition between *L* and *G* yields a polydomain structure indicative of epitaxial mismatches in domain orientations [152].

It is interesting to note that the *PL* phase was mostly detected in solutions of block copolymers. A PS-*b*-PI block copolymer solution with dialkyl phthalates formed the *G* phase via a *PL* intermediate when cooled from the disordered state. Nucleation and growth mechanisms of the various morphologies were detected by polarized optical microscopy. Temperature quenches allowed the investigation of the *T* dependence of the growth rate. The growth velocities, *v*, were in the range $0.06 \leq \mu m\ min^{-1} \leq 160$. In some samples the velocity increased with decreasing temperature (indicative of a thermodynamic driving force), whereas in other test specimens the velocity decreased (suggesting the influence of chain dynamics) [154]. The observed dependencies were in good agreement with the relations proposed by Goveas and Chastek, which predict the temperature dependence of velocity to be determined by (i) the difference between the interaction parameters at the quench temperature and the associated transition temperature, T_{OOD}, and (ii) the polymer chain relaxation times [154–156].

Using selective solvents for either PS or PI, different cubic spherical phases could be generated in solutions of a symmetric PS-*b*-PI diblock copolymer. Using shear, an *fcc* solution could be transformed into a mixture of *fcc* and *hcp*. Similarities to metals were found in the mechanisms present during the transitions from *fcc* or *hcp* lattices to the *bcc* lattice, when heating the system to the *bcc* region of the phase diagram in the absence of shear [157]. Very recently, the flow-induced transition from a *G* to a *C* morphology was reported in a PS-*b*-PB-*b*-PS triblock copolymer (SBS). Two-dimensional SAXS measurements revealed that the *G* structure transformed into uniaxially aligned cylinders with hexagonally close packing upon imposing the flow, where a plane of the hexagonally packed cylinders was preferentially aligned parallel to the substrate [158].

In the case of triblock terpolymers two different interfaces have to be considered for the morphological transitions, making the description of the transition process much more demanding. The transition from *L* to *G* morphology of a PI-*b*-PS-*b*-P2VP triblock terpolymer has been investigated by Suzuki et al. [159]. Directly after solvent casting the symmetric triblock ($\phi_{PI} = \phi_{P2VP} = 0.26$ and $\phi_{PS} = 0.46$) showed a metastable lamellar morphology (cf. also Sect. 2.2.12). Annealing the film for 14 h triggered the conversion to a *G*-type morphology taking more than 2 d to accomplish. It was reasoned that the lamellar structure could be formed because of a strong affinity of the P2VP to the polytetrafluorethylene material used in casting. The motif of layered lamellar structure (domain spacing 58 nm), however, guided the evolution of the *G* one, leading to its [110] orientation normal to the film surface and a repeating unit of 55 nm (Fig. 48).

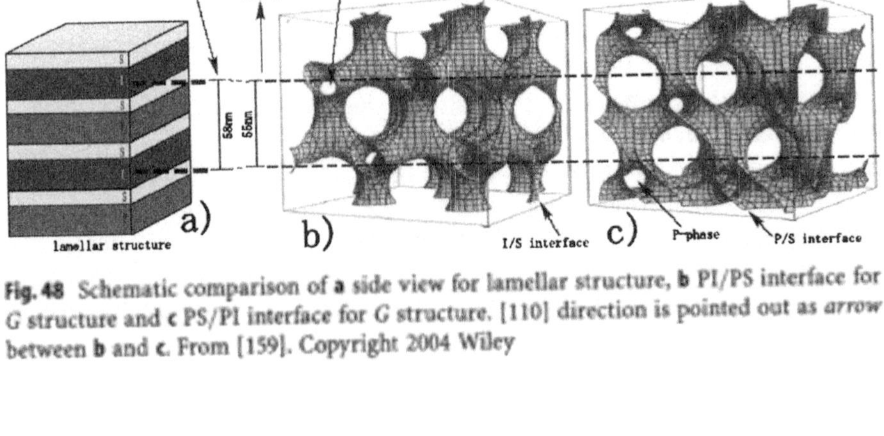

Fig. 48 Schematic comparison of **a** side view for lamellar structure, **b** PI/PS interface for *G* structure and **c** PS/PI interface for *G* structure. [110] direction is pointed out as *arrow* between **b** and **c**. From [159]. Copyright 2004 Wiley

4.4
Tuning Block Copolymer Morphologies Via Blending

In a strict sense, synthetic polymers are never one-component systems due to polydispersities in chain lengths, microstructures, tacticity, or composition (random copolymers, block copolymers). In the preceding sections blends of block copolymers with homopolymers were occasionally discussed. In this section we give an overview of various aspects of influencing block copolymer morphologies by blending. In the first part the influence of polydispersity is discussed. Then the swelling by low-molecular-weight components will be addressed. Examples of blending block copolymers with other polymers or block copolymers will follow, and finally the incorporation of block copolymers in other polymers by reactive blending will be briefly reviewed.

4.4.1
Effect of Molecular Weight Distribution

The effects of composition distribution on the morphology of PS-*b*-P2VP diblock copolymers were investigated by Matsushita et al. [160]. They produced PS-*b*-P2VP samples with various composition distributions but with constant average composition by blending. If the polydispersity indices of each block were lower than 1.7, the expected lamellar domains were detected (Fig. 49).

It is obvious from Fig. 49 that the domain spacing is not merely dependent on the number-average molecular weight of the constituent block copolymers. It is also governed by the condition for polymer segments to fill the 3D space constantly. Hence, a non-uniform compositional distribution also affects the spacing as schematically sketched in Fig. 50 for identical average chain length. In the non-uniform case (left), the segments for the longer chain compensates the shortage of the segments of the shorter ones lead-

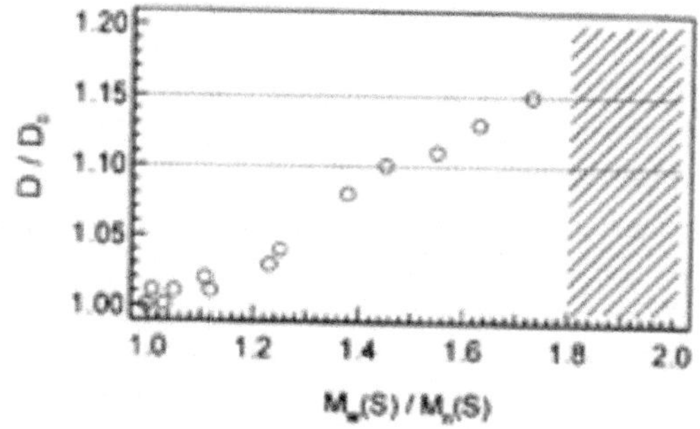

Fig. 49 Composition distribution dependence of reduced domain spacing, D/D_0, of PS-*b*-P2VP with single microphase-separated structure. $D_0 = 60.8$ nm is domain spacing of parent copolymer with $M_n = 125$ kg/mol. *Hatched region*: macrophase separation. From [160]. Copyright 2003 American Chemical Society

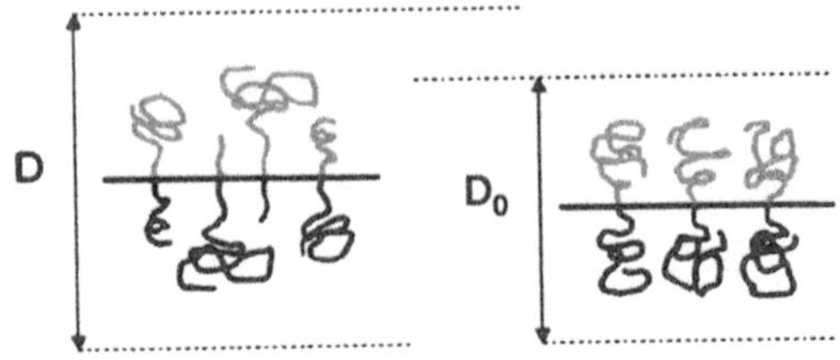

Fig. 50 Schematic comparison of microdomains formed by block copolymers with wide (*left*) and narrow (*right*) composition distributions. From [160]. Copyright 2003 American Chemical Society

ing to an increase in the domain spacing. Hence, the average surface area for a junction point of each block copolymer in blends with wide compositional distribution is smaller than that for regular monodisperse copolymers. As a result the spacing increases with increasing compositional distribution until macrophase separation takes place.

4.4.2 Effect of Low-Molecular-Weight Molecules

The addition of selective solvent into a block copolymer melt renormalizes the direct segment–segment interaction. Due to their different affinities to the solvent, the incompatibility of the blocks is enhanced. Both the polarity and the solubility of the segments are major factors determining an internal domain structure. While the highly diluted state leads to vesicular structures and micelles (see article "Block Copolymer Micelles" in this issue), we want to focus here on the concentrated states. A system studied in great detail is PS-*b*-PI in dialkyl phthalates. The dilution approximation for the segmental interaction parameter could be verified for OOTs with dioctyl phthalate as a non-selective solvent. Here, the polymer's volume fraction in solution, ϕ, can be described by a power-law $\chi \sim \phi^{-a}$ with $a = 1$. However, in case of an ODT, a ranges from 1.6 to 1.3. This is in disagreement with mean-field theory and may be attributed to the additional stabilization of disordered micelles [161]. Using selective solvents, the degree of swelling varies for the different blocks and results in a morphology change [162, 163]. The phase diagrams of an asymmetric PS-*b*-PI diblock in different solvents determined by SAXS is shown in Fig. 51. All phase transitions were reversible and showed some hysteresis on temperature scans.

Mean-field lattice theory proved to be capable of predicting the phase behaviour of the ternary block copolymer polyethylene-*b*-poly(propylene oxide)-*b*-poly(ethylene oxide), PE-*b*-PPO-*b*-PEO in the selective solvent water [164]. The ethylene block is known to be highly hydrophobic, and its hydrophobicity does not depend strongly on temperature. The difference in hydrophobicity between PPO and PEO, on the other hand, is moderate

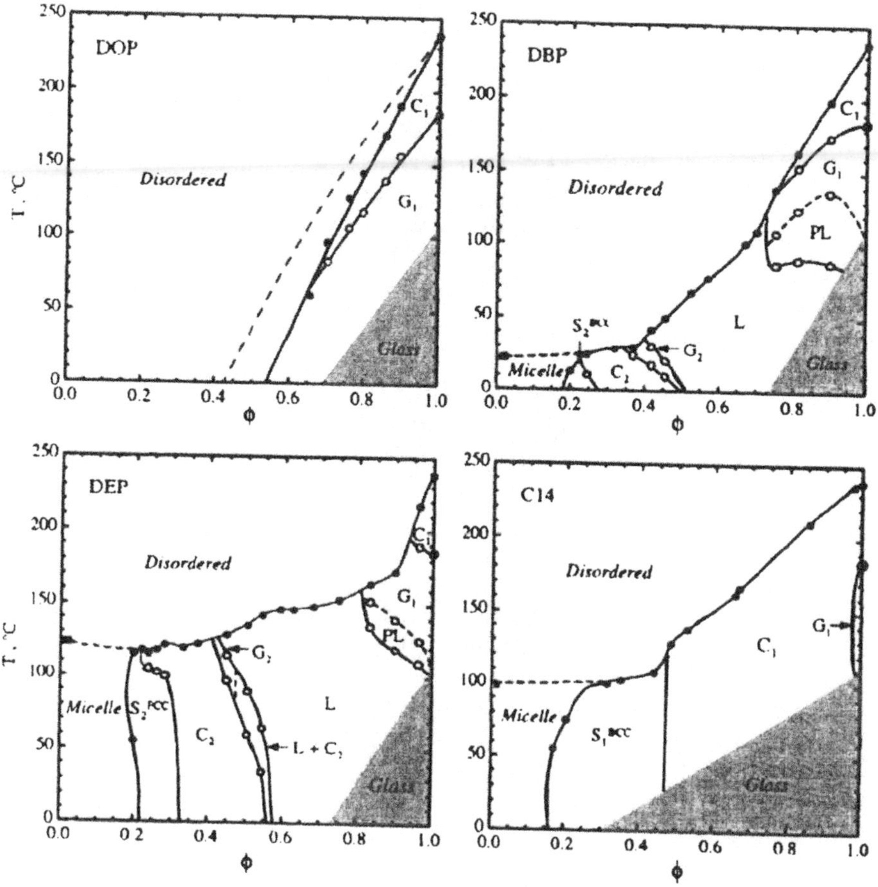

Fig. 51 Phase diagram for PS-PI diblock copolymer (M_n = 33 kg/mol, 31 vol% PS) as function of temperature, T, and polymer volume fraction, ϕ, for solutions in dioctyl phthalate (DOP), di-n-butyl phthalate (DBP), diethyl phthalate (DEP) and n-tetradecane (C14). (•) ODT; (○) OOT; (■) dilute solution critical micelle temperature, *cmt*. Subscript 1 identifies phase as "normal" (PS chains reside in minor domains); subscript 2 indicates "inverted" phases (PS chains located in major domains). Phase boundaries are drawn as guide to eye, except for DOP in which OOT and ODT phase boundaries (*solid lines*) show previously determined scaling of PS-PI interaction parameter ($\chi_{ODT} \sim \phi^{-1.4}$ and $\chi_{OOT} \sim \phi^{-1}$); *dashed line*: "dilution approximation" ($\chi_{ODT} \sim \phi^{-1}$). From [162]. Copyright 2000 American Chemical Society

(both are hydrophilic relative to the PE block). Nevertheless, it is known that PEO/PPO block copolymers form microphase-segregated structures in aqueous solution at sufficiently low temperatures. The combination of the two factors leads to an interesting composition-temperature behaviour of the system. In the temperature range (280–380 K) only four phases were found to be stable (Fig. 52): the hexagonal one, H_1, the lamellar one L_α, the reverse

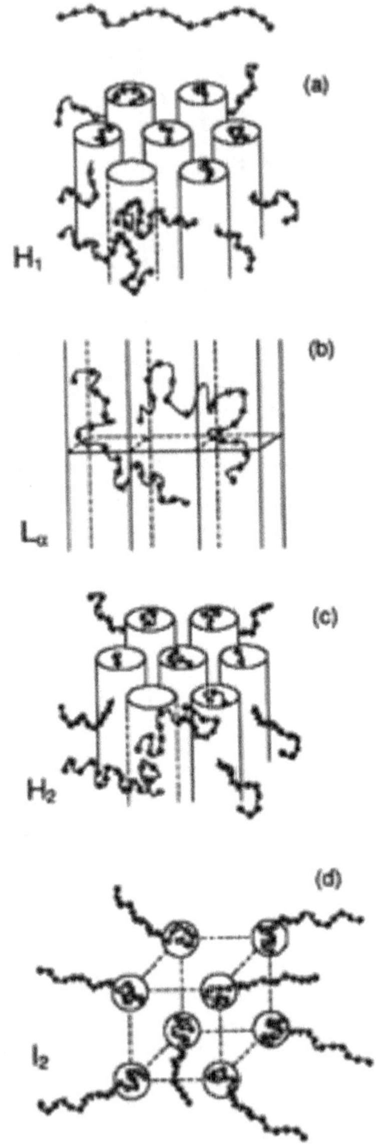

Fig. 52 Schematic illustration of different modes of self-organization of PE-*b*-PPO-*b*-PEO triblock terpolymer in water. •: PE segments, △: PPO segments, ○: PEO segments. Ordered phases are denoted as follows: **a** normal (water-continuous) hexagonal, H_1, **b** lamellar, L_α, **c** reverse (polymer-continuous) hexagonal, H_2 and **d** reverse micellar cubic, I_2. From [164]. Copyright 2003 EDP Sciences, Società Italiana di Fisica, Springer

hexagonal one, H_2, and the reverse cubic (discrete) one, I_2. At low temperatures, a transition from H_1 to L_α caused by increasing polymer concentration occurs, typical for many block copolymers composed of PEO and PPO.

A temperature-induced sequence of phase transitions $H_1 \rightarrow L_\alpha \rightarrow H_2 \rightarrow I_2$ over a wide range of polymer volume fractions is a new observation for the ABC triblock terpolymer systems. This thermotropic behaviour is caused by the strong temperature and concentration dependence of the PEO/water and PPO/water interactions. Elevated temperatures worsen the solubility of the PEO and PPO blocks and decrease the incompatibility between PE and PEO blocks, whereas the temperature effect becomes stronger for increasing polymer concentration. In the domain structure three layers, a PE-rich, a PPO-rich and a PEO-rich one, can be distinguished with the blocks forming the domains remaining unstretched. With increasing temperature the structure reverses from solvent-continuous to polymer-continuous and the size of the domain increases.

When used in a solution-casting process a selective solvent approach provides an effective tool for variation of bulk morphologies. The preparation of bulk samples by solution casting leads to a sequential microphase separation of the different copolymer blocks caused by the solvent's selectivity. The block with the lowest solubility separates first from the common solution. It self-assembles into microdomains containing less solvent than the remaining block(s). During the casting process the solvent concentration further decreases. In consequence, the remaining blocks start to segregate from each other because their repulsive interactions are no longer sufficiently screened by the solvent. This second segregation is geometrically controlled by the first one and may lead to unusual morphologies; e.g. Abetz et al. [165] employed this technique to obtain a polystyrene-*b*-polybutadiene-*b*-poly(methacrylic acid), PS-*b*-PB-*b*-PMAA, exhibiting a *hex* morphology by casting from tetrahydrofuran solution. The PMAA blocks form hexagonally packed cylinders, whereas the PS blocks are located in cylinders at the corners of the hexagon around each PMAA cylinder. The morphology's evolution commences with a self-assembly of the least soluble PMAA block into an array of hexagonally packed cylinders. At this concentration the PS and PB blocks are still not segregated from each other. Further evaporation of the solvent leads to a separation of PS and PB blocks, where PS organizes into the second set of cylinders as already described before. In this pattern the PB chains suffer the lowest frustration. Unlike the PS-*b*-PB-*b*-PMAA system the structure evolution of a polystyrene-*b*-poly(ethylene-*co*-butylene)-*b*-poly(methyl methacrylate), PS-*b*-PEB-*b*-PMMA, commences in chloroform [166, 167] with the organization of the PEB middle block in *hex* cylinders. Depending on the orientations of the PS and PMMA end blocks, different final structures may result after the solvent is completely removed and the PS and PMMA blocks are microphase separated (Fig. 53). This leads to the so-called "knitting pattern" morphology, which was first described by Stadler's group [168, 169].

In principle, any kind of low-molecular-weight species miscible with at least one component of the block copolymer will alter the phase space. A re-

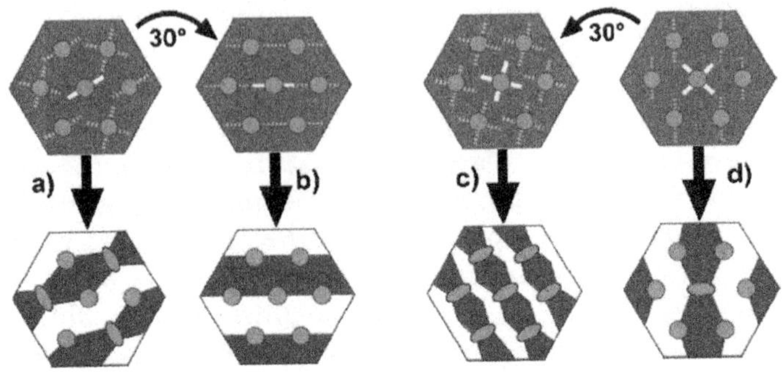

Fig. 53 Different possible scenarios during formation of knitting pattern morphology. *Upper row*: PEB is microphase separated, PS and PMMA are still dissolved in chloroform; *lower row*: final morphology after removal of solvent. From [166]. Copyright 2001 American Chemical Society

cent example is the doping of the PEO block of a PI-*b*-PS-*b*-PEO or PS-*b*-PI-*b*-PEO copolymer with lithium perchlorate, as shown in Figs. 54 and 55. In the PS-*b*-PI-*b*-PEO system (Fig. 54), the core-shell gyroid (CSG) and semiperforated lamellae (SPL) morphology are lost, the two-domain and three-domain lamellae (LAM_2, LAM_3), and the pillared lamellar structure (PLS) are more or less preserved and the core-shell cylinder (CSC) geometry is expanded [170].

Doping the PEO phase with lithium perchlorate in the PI-*b*-PS-*b*-PEO (Fig. 55) does not lead to a loss of morphologies. However, the orthorhombic phase (cf. Fig. 18 for a more extensive discussion) changes into a core-shell cylinder morphology [171].

The loss of phase complexity in both systems may be attributed to an increase of the PS/PEO and PI/PEO interaction parameters. Because $LiClO_4$ is selectively located in the PEO domains, the interaction parameters ($\chi_{PS\text{-}PEO}$ and $\chi_{PI\text{-}PEO}$) must increase, leading to variations in domain type and dimension. As the lithium salt increases the polarity (and presumably the solubility parameters) of the PEO domains, the interfacial tensions between PEO and PI, and PEO and PS are elevated. Thus, a reduction in the overall PEO interfacial area is required, which necessitates additional chain stretching. In consequence, the CSC structure becomes dominant when comparing doped and non-doped samples [171] (Figs. 54 and 55b).

4.4.3
Binary Blending with Homopolymer (AB + A)

The effect of blending an AB diblock copolymer with an A-type homopolymer has been the subject of many research activities. On a theoretical basis the subject was investigated e.g. by Whitmore and Noolandi [172] and Matsen [173]. If a diblock exhibiting lamellae morphology is blended with a homopolymer of high molecular weight, macrophase separation between the

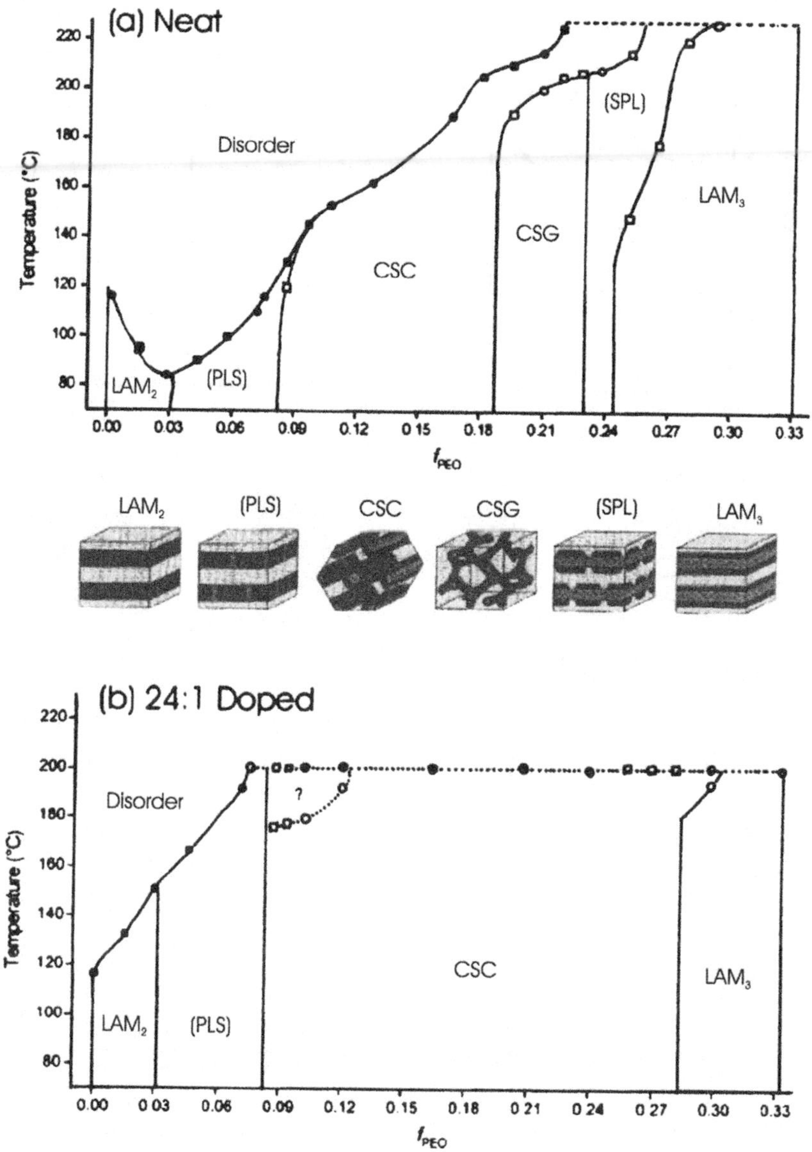

Fig. 54 Phase diagram along $\phi_{PS} = \phi_{PI}$ isopleth detailing structures found in PS-*b*-PI-*b*-PEO triblock terpolymer system with ODTs (•, ■) and OOTs (○, □) for triblock samples (•, ○) and triblock blends (■, □). **a** Neat phase diagram. *Dashed line*: experimental ceiling at 225 °C. **b** Phase diagrams detailing structures observed in lithium-doped PS-*b*-PI-*b*-PEO triblock isopleth system. ⊙: Doped triblock samples; ⊡: doped triblock blends at experimental ceiling at 200 °C (*dashed line*). Morphologies sketched between **a** and **b**. LAM$_2$: two-domain lamellar structure; PLS: pillared lamellar structure; CSC: core-shell cylinder; CSG: core-shell gyroid; SPL: semiperforated lamellae; LAM$_3$: three-domain lamellae. From [170]. Copyright 2002 American Chemical Society

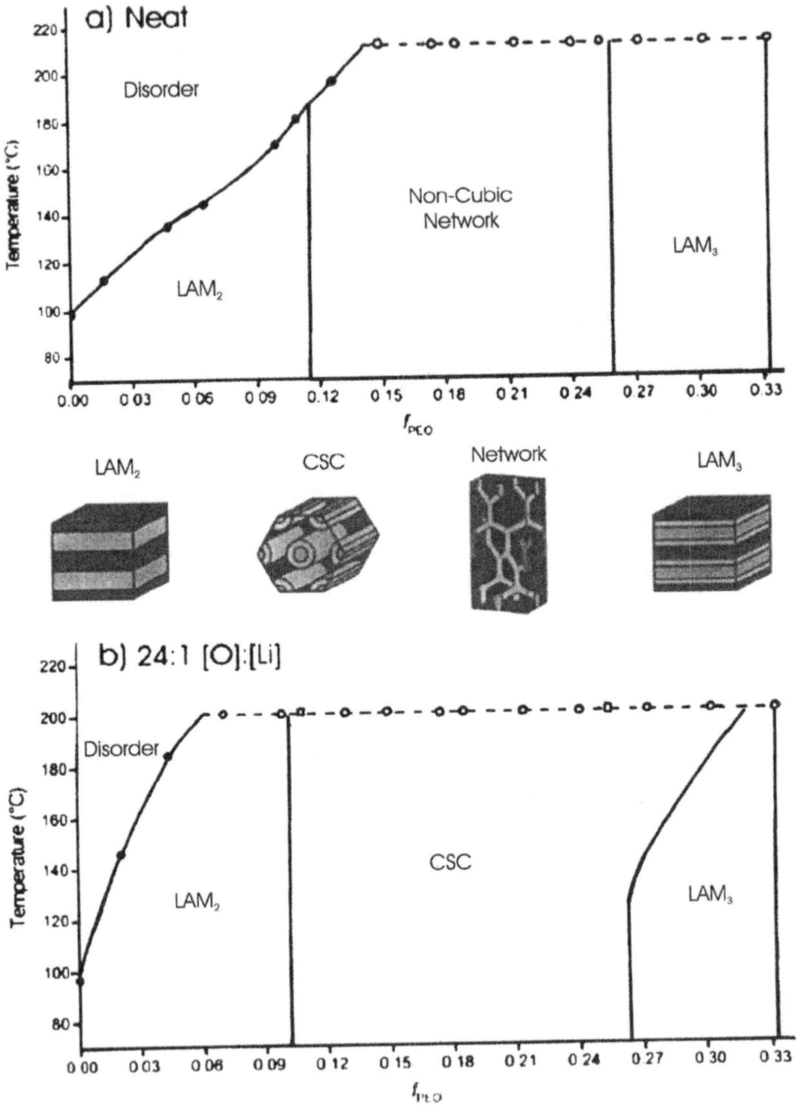

Fig. 55 Phase diagram along $\phi_{PS} = \phi_{PI}$ isopleth detailing structures found in PI-*b*-PS-*b*-PEO triblock terpolymer system with ODTs (•, ■) and OOTs (○, □) for triblock samples (•, ○) and triblock blends (■, □) and experimental ceiling at 225 °C (*dashed line*). **a** Neat phase diagram. One non-cubic triply periodic phase lies between two-domain and three-domain lamellae, tentatively identified as single-network structure with *Fddd* space group symmetry as illustrated in sketch and in Fig. 19. **b** Phase diagrams detailing structures observed in lithium-doped PS-*b*-PI-*b*-PEO triblock isopleth system. Morphologies are sketched between **a** and **b**. LAM$_2$: two-domain lamellae structure; CSC: core-shell cylinders; LAM$_3$: three-domain lamellae. From [171]. Copyright 2003 American Chemical Society

diblock-rich lamellar phase and a homopolymer-rich disordered phase is observed which is caused by an attractive interaction between diblock bilayers. In contrast, a low-molecular-weight homopolymer leads to a repulsive interaction between bilayers. Here, the homopolymer may be added to the lamellar phase indefinitely without macrophase separation. The A blocks of the block copolymer are concentrated near the interface, while the A homopolymer is preferentially located in the central region of the A-rich domains. Tension in the A blocks pushes the homopolymer toward the middle of the A-rich domains. Entropy of mixing counters this effect and favours a more uniform homopolymer distribution. The latter effect is enhanced by lowering the molecular weight of the homopolymer. To accommodate homopolymer near the interface, the microstructure may undergo a phase transition to a geometry where the interface is curved less toward the A-rich domain. Even if this effect is not strong enough to cause a phase transition, homopolymer may penetrate toward the interface by spreading apart the diblocks and in consequence increasing the interfacial area. This allows the B blocks to contact, leading to a reduced thickness of the B-rich domains. In summary, the studies predicted that blending a diblock copolymer with a homopolymer would result in a shift of phase boundaries and in the stabilization of new morphologies.

A phase diagram of the symmetric PS-*b*-PI blended with PS homopolymer of shorter chain lengths was constructed by Bodycomb et al. [174]. The effect of blend composition on the ODT is shown in Fig. 56 along with the results of mean-field calculations. In analogy to MFT the addition of homopolymer decreases the ODT temperature for the nearly symmetric diblock copolymer.

The phase diagram in Fig. 56 reveals an irreversible temperature-induced OOT between *C* and *G*, with the former being the low-temperature phase, which differs from previous studies on PS-*b*-PI/PS or PI blends [10]. The irreversibility of the OOT was attributed to a kinetic effect. At temperatures between ODT and OOT the system assembles in a *G* morphology. This structure is kinetically trapped and persists at temperatures below OOT, where cylinders are expected to be formed.

Recently, the effect of blending the commercially available PI-*b*-PS-*b*-PI triblock copolymer VECTOR 4111 with PS and PI on the phase behaviour was investigated [175]. Addition of low-molecular-weight homopolymers to an asymmetric triblock copolymer caused contraction of the domain spacing. The addition of majority-component PI triggered the transition from a cylindrical to a spherical microdomain structure. This indicates that the increase in curvature is favoured compared either to a stretching of the PI block or a contraction of the PS block because it involves less conformational entropy loss. The addition of a small concentration of minority component PS leads to an intermediate structure, a possible precursor to a lamellar phase. Addition of a 1 : 1 blend of PI and PS changed the cylindrical morphology neither in bulk nor in thin films and led to enhanced alignment of cylin-

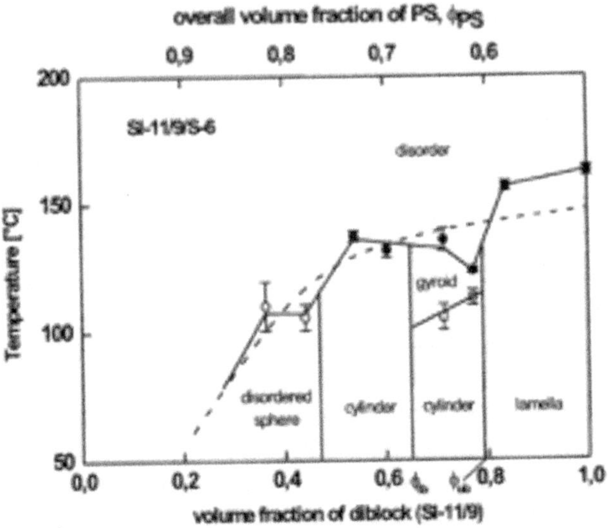

Fig. 56 Phase diagram of blend of PS-*b*-PI with PS. •: T_{ODT}, ○: T_{DMT}, □: T_{OOT}. *Vertical lines* separating microdomain structures are obtained from total volume fraction PS in system. *Dashed line*: results of mean-field calculation for ODT. The OOT line which exists at volume fractions $\phi_{lb} \leq \phi_{PS} \leq \phi_{ub}$ was obtained during a heating process. From [174]. Copyright 2000 American Chemical Society

ders parallel to the surface in case of thin films. This behaviour could have been caused either by the segregation of PI to the substrate, resulting in an additional "wetting layer" or by a "plasticization" of the PS by low molecular weight homopolymer. However, at one particular blend composition the ODT dropped enormously. A similar behaviour was reported for pure block copolymers [176] in blends of PS with polystyrene-*b*-poly(2-vinylpyridine) and with PS-*b*-PI copolymers [11].

4.4.4
Blending a Triblock Terpolymer with Two Homopolymers (ABC + A and C)

As in the case of diblock polymers, the morphology of triblocks can be altered by blending homopolymers with them. Unlike the case of a binary ABA triblock, where the middle block possesses either a bridge or a loop conformation (Fig. 57b), the midblock of a ternary ABC triblock with incompatible A and C blocks will only form bridges (Fig. 57a). In consequence, adding B homopolymer to the ternary triblocks results in swelling of the B domain. This leads inevitably to an elongation of the bridged midblock B chains. In contrast, the addition of A and C homopolymer does not require elongation of the triblock's A- and C-block chains, and in consequence the conformation of the A- and C-block chains should not be much affected [159].

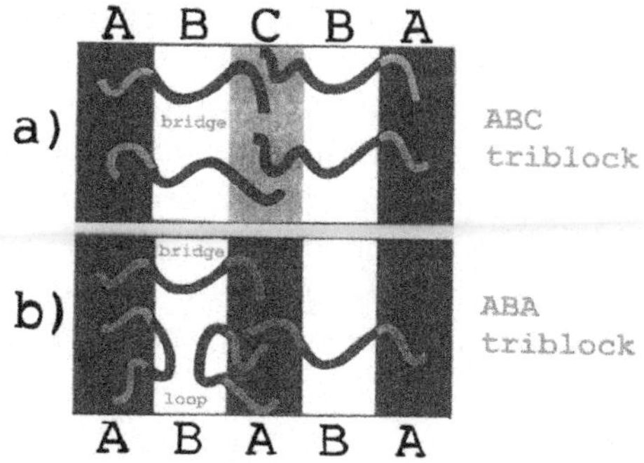

Fig. 57 Schematic comparison of chain conformations of the midblock for ABC and ABA triblocks. ABC triblock terpolymers (**a**) have bridge conformations only, whereas ABA triblock copolymers (**b**) have bridge and loop conformations. From [159]. Copyright 2002 Wiley

It is suggested that in the case of ABC triblock terpolymers the gyroid phase extends into a stronger segregation region [78]. However, the composition window for that phase is quite narrow. Sugiyama et al. [177] tried to enlarge that window by blending PS-*b*-PI-*b*-PDMS triblock copolymer ($\phi_{PS} = 0.20$; $\phi_{PI} = 0.59$; and $\phi_{PDMS} = 0.21$; $M_n = 40\,000$) with equal amounts of low-molecular-weight PS ($M_n = 2360$) and PDMS ($M_n = 2200$). It was found that even small amounts of PS and PDMS homopolymers create a CSG structure in a remarkably wide range. As the homopolymer volume fraction ϕ_H exceeds 0.45, the morphology changes to hexagonally packed CSCs before phase separation occurs at $\phi_H \approx 0.65$.

The appearance and persistence of core-shell structures as well as the occurrence of phase separation are attributed to a small asymmetry in the χ -parameters ($\chi_{PS\text{-}PI} = 0.06$, $\chi_{PI\text{-}PDMS} = 0.09$ and $\chi_{PS\text{-}PDMS} = 0.20$). Hence, a PDMS core surrounded by a PI shell embedded in a PS matrix results in a smaller inner diameter interfacial area, relative to that for the PS-PI case. In a blend of a PS-*b*-PI-*b*-PDMS triblock with PS and PDMS homopolymers, more PS homopolymer is expected to be found in the corona of the PS block than PDMS homopolymer in the corona of the PDMS block because the penalty for contact between the PI block and PDMS homopolymer is larger. In consequence, the distribution of homopolymers favours an expanded PS-PI interface, making the core-shell morphologies, gyroid and cylinder, more prevalent.

Addition of the middle B block to an ABC triblock terpolymer has been investigated by Suzuki et al. for the PI-*b*-PS-*b*-P2VP system [159]. Starting from the lamellar structure of the unblended triblock ($\phi_{PS} = 0.42$) PS homopolymer was subsequently added. At $\phi_{PS} \approx 0.50$ a morphological transformation into a gryoid structure is observed. Even if the volume fraction of PS is increased up to $\phi_{PS} = 0.60$, the cell size of the gyroid structure will remain constant (Fig. 58a, blend I). Based on a triblock terpolymer exhibiting gyroid morphology in the unblended state, the volume fraction of PS was increased up to $\phi_{PS} = 0.7$ resulting in a cylindrical structure (Fig. 58a, blend II). Those results imply that PS chains of the triblock can be stretched by adding PS homopolymer; however, they reach their limits for chain stretching eventually. This limit of chain stretching for the PS block may be caused by conformational frustration as the middle block possesses a bridge conformation and both ends have to be tethered on different interfaces. The situation changes

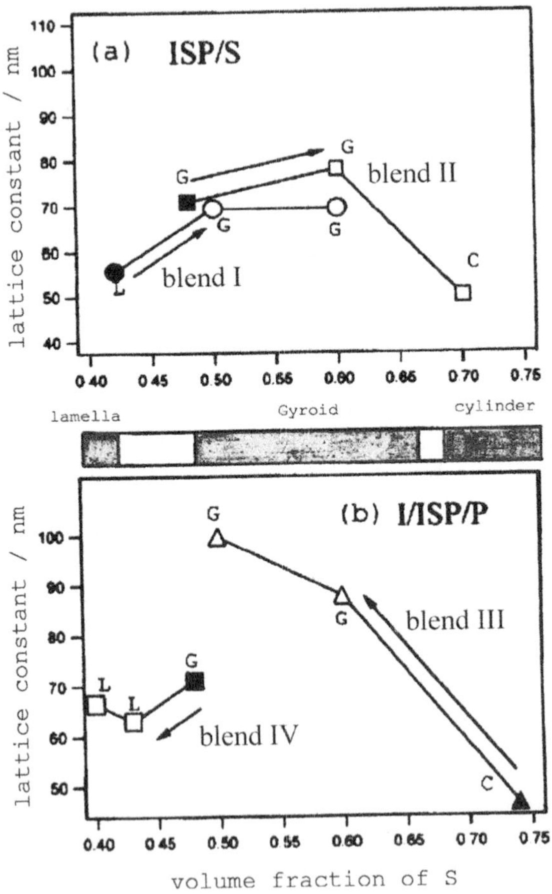

Fig. 58 Lattice constants vs. volume fractions of PS phase for **a** blending a PI-*b*-PS-*b*-P2VP triblock terpolymer with PS homopolymer (blends I and II) and **b** blending a PI-*b*-PS-*b*-P2VP triblock terpolymer with PI and P2VP homopolymers (blends III and IV). *Arrows*: variations of ϕ_{PS} with increasing volume fractions of added homopolymers. •, ■, ▲: lattice constants of pure triblock terpolymers; ○, □, △: lattice constants of blends. *Gray band* between **a** and **b** expresses experimentally obtained microphase separation phase diagram for unblendend PI-*b*-PS-*b*-P2VP. From [159]. Copyright 2002 Wiley

if the PI-*b*-PS-*b*-P2VP block copolymer is blended with PI and P2VP homopolymers simultaneously (Fig. 58b). Starting from a cylindrical structure with PI and P2VP cylinders tetragonally arranged in a PS matrix ($\phi_{PS} = 0.74$), blending leads to a gyroid structure with lattice constants increasing with decreasing ϕ_{PS} (Fig. 58b, blend III). The same trend is followed for lower volume fractions of PS, where the transformation to a lamellar morphology is observed for $\phi_{PS} < 0.45$ (Fig. 58b, blend IV). Here, the PI and P2VP domains have no bridge conformations, and so PI and P2VP homopolymers can be localized at the centre of PI and P2VP domains without affecting significantly the conformation of the respective block chains. Thus, the addition of A and C homopolymers to ABC triblock terpolymers may contribute slightly to the conformational entropy of block chains in forming a microphase-separated structure.

4.4.5
Blending of Similar Block Copolymer Types (AB + A'B')

The phase behaviour of a blend of two PS-*b*-PI copolymers depending on the components' molecular weight and their ratio was investigated by Yamaguchi et al. [178–181]. Both copolymers were of nearly symmetric composition; however, they differed in their molecular weights. For a given molecular weight ratio of the two constituent PS-*b*-PIs the parameter space of temperature and blend composition is depicted in Fig. 59.

An ODT was found for mixtures whose weight fractions of the higher molecular weight copolymer ranged from $0 \leq \phi \leq 0.2$. Contrary to the results of previous studies [182, 183], the T_{ODT} was the highest for the short copolymer and decreased when the fraction of higher molecular weight copolymer was increased. A small fraction of the latter block copolymer may be incorporated into the lamellae of the shorter one, leading to a destabilization of lamellar packing. The instability results in a decrease of T_{ODT} for the blends. For a weight fraction of $\phi = 0.2$, a hysteresis between heating and cooling processes was observed. Non-uniform fluctuation effects after phase separation may account for this obeservation. For the blends of $0.2 \leq \phi \leq 0.3$, a partial miscibility of both components could be observed and a macroscopically phase-separated system with two coexisting ordered phases was obtained. This is qualitatively consistent with the predictions of self-consistent-field theory (SCFT) [184]. In this region T_{ODT} increased abruptly and finally exceeded the decomposition temperature of the polymer (i.e. > 200 °C).

For blend ratios of $0.2 \leq \phi \leq 0.4$ a morphology of PS cylinders was obtained, although both block copolymers self-assemble on their own into lamellar patterns. Interestingly, non-lamellar morphologies tend to appear when small amounts of higher molecular weight PS-*b*-PI are added to the matrix polymer of low molecular weight. These morphologies persist up to a ratio of $\phi = 0.50$. On the other hand, when the higher molecular

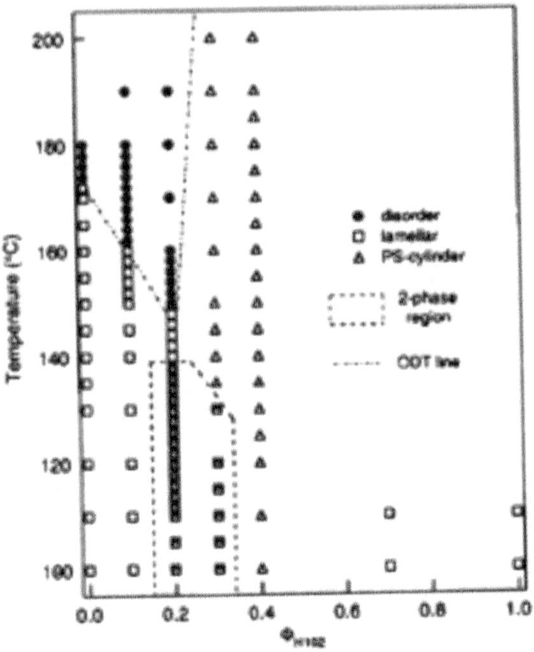

Fig. 59 Phase diagram for blend consisting of two symmetric PS-*b*-PI block copolymers of different molecular weights in parameter space of temperature and fraction of higher molecular weight copolymer, ϕ. •: disordered state; □: lamella; △: PS cylinder. From [179]. Copyright 2001 American Chemical Society

weight PS-*b*-PI is the major component, the lamellar morphology is retained. This finding was attributed to the slightly asymmetric compositions of the low-molecular-weight PS-*b*-PI leading to a non-zero spontaneous curvature (Fig. 60a). Due to the low molecular weight the polymer does not self-assemble into a cylindrical morphology. If cylinders are formed, their radii are too small compared with those expected from the spontaneous curvature (broken lines in Fig. 60b). In consequence, lamellae are formed exhibiting a vanishing spontaneous curvature. If miscible PS-*b*-PI chains of higher molecular weight are added, the central part of the cylinder domains is filled and the cylinders with a mean curvature close to the spontaneous curvature inherent in the low-molecular-weight PS-*b*-PI will be realized as sketched in Fig. 60c.

In general, the less the asymmetry in the composition of the low molecular weight PS-*b*-PI, the larger molecular weight of the other component is required to allow formation of cylinders. This trend leads, however, to a certain ratio in which the system macrophase separates before cylinders are formed. Therefore, the formation of cylinders tends to occur in a range in which the miscibility between the chains abruptly changes.

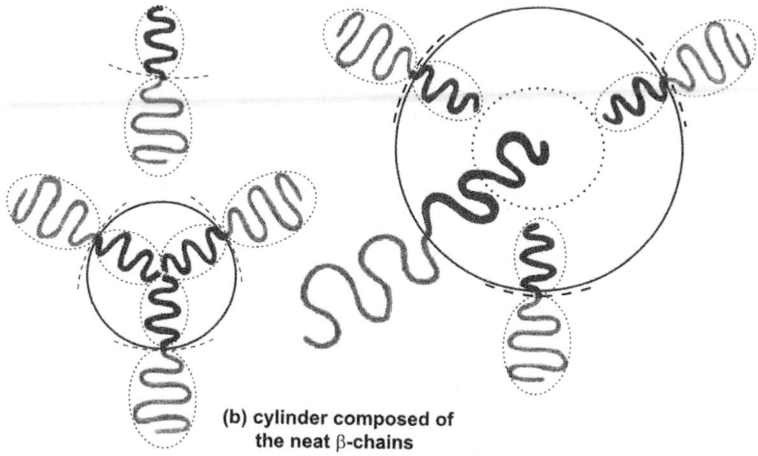

Fig. 60 Schematic illustration for formation of cylindrical morphology in a blend of slightly asymmetric lower molecular weight PS-*b*-PI (β-chain) with large symmetric PS-*b*-PI (α-chain). **a** Molecule of β-chain with non-zero spontaneous curvature. **b** Cylindrical morphology formed by neat β chains shown in **a**. Here mean curvature of cylinder (*solid line*) is larger than spontaneous curvature of β-chain (*dashed lines*). **c** Cylindrical morphology formed by binary blend of β-chains shown in a and large symmetric copolymers (α-chain). In this case, mean curvature of cylinder closely fits to spontaneous curvature of β-chain. From [180]. Copyright 2001 American Chemical Society

The effect of mixing two PS-*b*-PI block copolymers exhibiting a similar overall molecular weight ($M_n \approx 1.25 \times 10^4$) but with complementary compositions, $\phi_{PS} = 0.81$ and $\phi_{PS} = 0.21$, on the phase behaviour was also investigated [181]. Both neat diblocks are in disordered state at all temperatures. However, the 50/50 (wt %/wt %) blend of both polymers behaved as if they were the LCST (lower critical solution temperature) type of polymer blend as schematically highlighted in Fig. 61a. In this type of blend the constituent copolymers were miscible at the temperatures below LCST (cf. Fig. 61b). For temperatures exceeding LCST macroscopic phase separation into the two disordered phases will be observed, as shown schematically in Fig. 61c. The miscibility at lower temperatures originates from the local segregation of the repulsive PS and PI segments into the PS and PI microdomains (Fig. 61b, inset). This effect is mainly driven by the increased interaction parameter between PS and PI segments with decreasing temperature. As this pseudo-LCST type phase stems from the particular PS-*b*-PI system, it might not necessarily be transferred to other blends in general. Nevertheless, a similar behaviour was anticipated from SCF calculations [185].

Court and Hashimoto [186] investigated morphological transitions triggered by blending two block copolymers with different volume fractions.

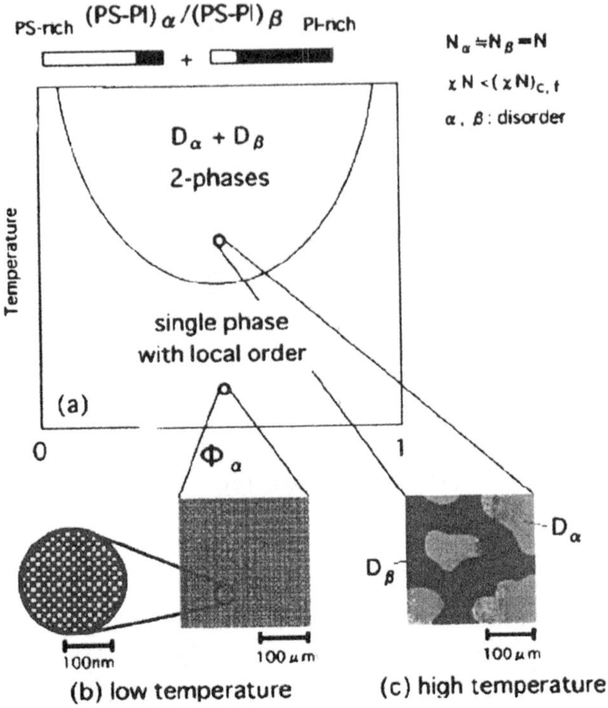

Fig. 61 **a** Schematic representation of phase diagram of blend of PS-rich (α) with PI-rich (β) PS-*b*-PI block copolymer in parameter space of ϕ_α, and T. Expected morphologies of blend specimen are also sketched at **b** low and **c** high temperatures. Note phase diagram is effective only for $\Phi_\alpha = 0.5$. From [181]. Copyright 2001 American Chemical Society

A shift of phase boundaries between neighbouring morphologies was observed to be dependent on the ratio (Fig. 62). These blends consist of an asymmetric PS-*b*-PI diblock, *as*, exhibiting a *bcc* morphology of PS spheres in a PI matrix and symmetric three PS-*b*-PI diblocks of lower molecular weight, s_i. Two of those shorter ones exhibited a lamellar morpology, while one of those was in the disordered state. The blend's phase diagram was not simply dependent on the overall volume fraction of PS and PI; it was strongly affected by the chain length ratio, r, of the asymmetric and the symmetric diblock.

The phase diagram of the blends in the parameter space of r and ϕ_{PS} at constant temperature is shown in Fig. 62. The blend morphologies are the classical ones of pure AB diblocks (Fig. 1). They are found in the same order, ranging from *L* via *C* to, finally, *S*, for decreasing the total volume fraction of PS (ϕ_{PS}). The shift in the composition range enlarges as the difference in molecular weight between the two diblocks increases. Therefore, blending diblocks of different molecular weights and compositions indicates a promis-

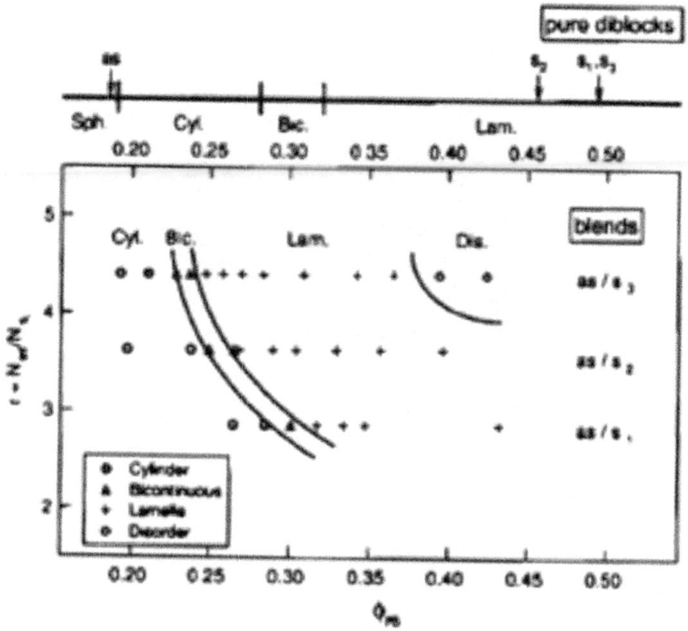

Fig. 62 Morphology variation of blends of two diblock copolymers as/s_i ($i = 1, 2, 3$) in parameter space of chain length ratio r and ϕ_{PS}. *Upper line*: Morphologies of asymmetric (as) and three symmetric (s_1, s_2 and s_3) PS-*b*-PI copolymers. From [186]. Copyright 2001 American Chemical Society

ing route to break the interdependence between morphology and volume fractions.

4.4.6
Blending of Triblock (ABA or ABC) and Diblock (AB or AC) Copolymers

The effect of blending a lamellar triblock PS-*b*-PI-*b*-PS copolymer with an *L* or C-type PS-*b*-PI copolymer can either increase or decrease the long period of the parent block copolymer. Generally speaking, PS-*b*-PI-*b*-PS/PS-*b*-PI blends composed of PS-*b*-PI chains with relatively short PI blocks tend to reduce the thickness of the PI lamellae. Here, the PI blocks from the PS-*b*-PI copolymer fill the space close to the interface. Such block arrangement forms a bi-disperse polymer layer, forcing the PI blocks from the PS-*b*-PI-*b*-PS copolymer to reside near the lamellar midplane and promoting the formation of midblock bridges [187].

Some work has been done on blends of ABC and AB or AC or AB′C block copolymers, such as polystyrene-*b*-polybutadiene-*b*-poly(methyl methacrylate) (PS-*b*-PB-*b*-PMMA) triblock terpolymers with PS-*b*-PB or PB-*b*-PMMA or other systems. Besides known morphologies for these block copolymers (though at other overall compositions with respect to the different chemical

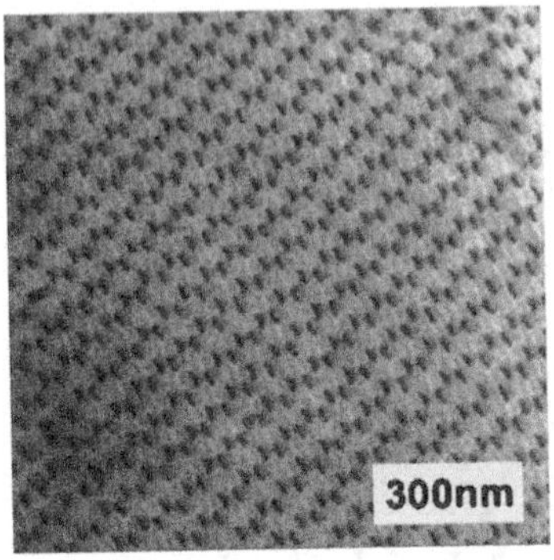

Fig. 63 TEM micrograph of blend of PB-*b*-PS-*b*-P2VP with PS-*b*-P2VP stained with OsO_4 and CH_3I. *Grey*: P2VP domains, *dark*: PB domains, *white*: PS domains. From [192]. Copyright 2003 Wiley

repeating units), also morphologies could be obtained which do not exist in the pure block copolymers. These works have been reviewed [5, 8]. In particular, a non-centrosymmetric lamellar superstructure was found in various blends of ABC- and AC-block copolymers [188–190] and examined theoretically using self-consistent mean-field theory [191]. An example of another unusual superstructure is shown in Fig. 63 for a blend of lamellar PB-*b*-PS-*b*-P2VP and PS-*b*-P2VP [192].

4.4.7
Blending of ABC Miktoarm-Star Terpolymers with AB-Diblock Copolymers

Besides linear-block copolymers, also blends with miktoarm-star terpolymers have been reported [193]. Blending a PS-*arm*-PB-*arm*-P2VP miktoarm-star terpolymer showing hexagonal symmetry with a PS-*b*-P2VP diblock

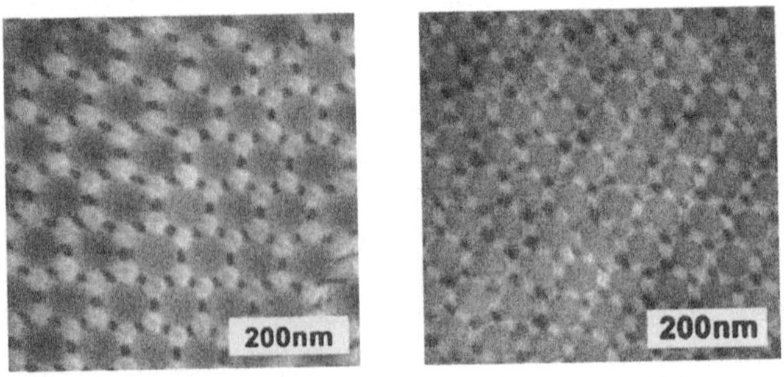

Fig. 64 TEM micrographs of pure PS-*arm*-PB-*arm*-P2VP (*left*) and blended with PS-*b*-P2VP (*right*). Samples are stained with OsO_4 and CH_3I. *Grey*: P2VP domains, *dark*: PB domains, *white*: PS domains. From [193]. Copyright 2004 e-polymers

copolymer leads to a superstructure in which the PS and PB domains (both being composed of the corresponding blocks of the two blend partners) exchange their symmetry positions, as demonstrated in Fig. 64 [193].

A PS-*arm*-PB-*arm*-P2VP showing lamellar morphology can form a CSG or CSC morphology when being blended with PS-*b*-P2VP or poly(2-vinyl pyridine)-*b*-poly(cyclohexyl methacrylate), P2VP-*b*-PCHMA respectively. In both cases the P2VP domains form the matrix, and in the latter the attractive interactions between PCHMA and PS lead to a stronger interdigitation of these blocks, resulting in a stronger interfacial curvature (cylinders instead of gyroid) [193].

4.4.8
Blending of Block Copolymers with Hydrogen Bonding Interactions

As an example of blends with attractive interactions, Fig. 65 shows a superstructure in which interactions between methacrylic acid groups and pyridine side groups of a polystyrene-*b*-polybutadiene-*b*-poly(*t*-butyl methacrylate-*stat*-methacrylic acid) (PS-*b*-PB-*b*-P(MAA-*stat*-*t*BMA)) triblock quaterpolymer and a PS-*b*-P2VP diblock copolymer lead to a wavy lamellar structure with cylinders from mixed P2VP and P(MAA-*stat*-*t*BMA) blocks [194].

Attractive interactions are also the reason for the self-assembly of PS-*b*-PB-*b*-PMMA at the interface of poly(styrene-*co*-acrylonitrile), SAN, and poly(2,6-dimethylphenylene ether), PPE. In this blend, PS and PPE are miscible on one side and PMMA and SAN are miscible on the other one, with negative χ parameters. This blend, in which the rubbery domain is located at the interface between SAN/PMMA and PPE/PS, was originally prepared by coprecipitation of all components from a common solution [195]. From a processing point of view, in this system the difficulty was to get the dispersion of PPE in SAN via melt mixing of SAN, PPE and the triblock terpolymer.

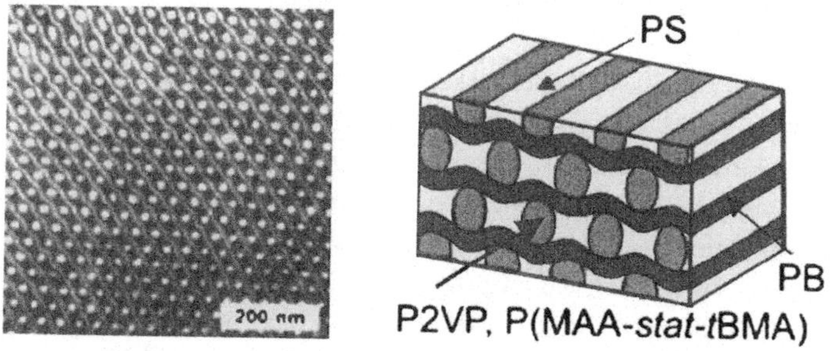

Fig. 65 TEM micrograph of PS-*b*-PB-*b*-P(MAA-*stat*-*t*BMA) blended with PS-*b*-P2VP (stained with RuO_4); *right*: schematics. *Grey*: mixed P(MAA-*stat*-*t*BMA) and P2VP domains, *dark*: PB domains, *white*: PS domains. From [194]. Copyright 2003 American Chemical Society

Besides the thermodynamic properties, viscosity effects also played an important role, as in all commercial blends prepared by a melt process [196].

4.5 Reactive Blending with Block Copolymers

4.5.1 Poly(ethylene oxide)-*b*-poly(ethylene-*alt*-propylene)

The phase space of poly(ethylene oxide)-*b*-poly(ethylene-*alt*-propylene), PEO-*b*-PEP, block copolymers with 50 or 26 vol % PEO blended with various amounts of a commercial epoxy resin (cured with 4,4′-methylenedianiline, MDA, after blending) were explored by Dean et al. [197]. As the epoxy is miscible with PEO [198, 199] but not with PEP, the phase behaviour (Fig. 66) follows a sequence of transitions associated with epoxy solvation of the PEO brush. Order-to-order phase transitions occur to accommodate the enlarged PEO/epoxy domain as the epoxy concentration in the blend increases (Fig. 66a). Unblended symmetric PEO-*b*-PEP self-assembles into lamellae that can take up to 25 wt % of epoxy before reorganizing into a *G* phase. The latter can be observed for blending ratios up to 40 wt % epoxy. Hexagonally packed cylinders of PEP in a matrix of PEO and epoxy are formed in blends containing between 40 and 60 wt % epoxy. The *S* lattice occurs above 60 wt % epoxy. In summary, curvature increases with increased epoxy concentration to accommodate the enlarged PEO/epoxy domain [200]. The undiluted asymmetric PEO-*b*-PEP block copolymer self-assembles into hexagonally packed cylinders of PEO in a matrix of PEP. The addition of epoxy to the minority cylinder phase enlarges the cylinder radius, reducing its curvature. The *C* phase transforms into a *G* phase with PEO/epoxy channels near 10 wt % epoxy. Lamellae are formed at ca. 20 wt % epoxy (Fig. 66b).

When novolac epoxy resin, for instance, was blended with nearly symmetric poly(methyl acrylate-*co*-glycidylmethacrylate)-*b*-polyisoprene and cured with MDA, a significantly different phase behaviour was found [201] (Fig. 67).

Instead of the familiar sequence of morphologies, a broad multiphase window centred at relatively high concentrations (ca. 50–70% block copolymer) truncates the ordered lamellar regime. At higher epoxy concentrations worm-like micelles and eventually vesicles at the lowest compositions are observed. Worm-like micelles are found over a broad composition range (Fig. 67). This morphology is rare in block copolymer/homopolymer blends [202] but is commonly encountered in the case of surfactant solutions [203] and mixtures of block copolymers with water and other low molecular weight diluents [204, 205].

Rebizant et al. [206, 207] investigated polystyrene-*b*-polybutadiene-*b*-poly(methylmethacrylate-*co*-glycidylmethacrylate) triblock quaterpolymers

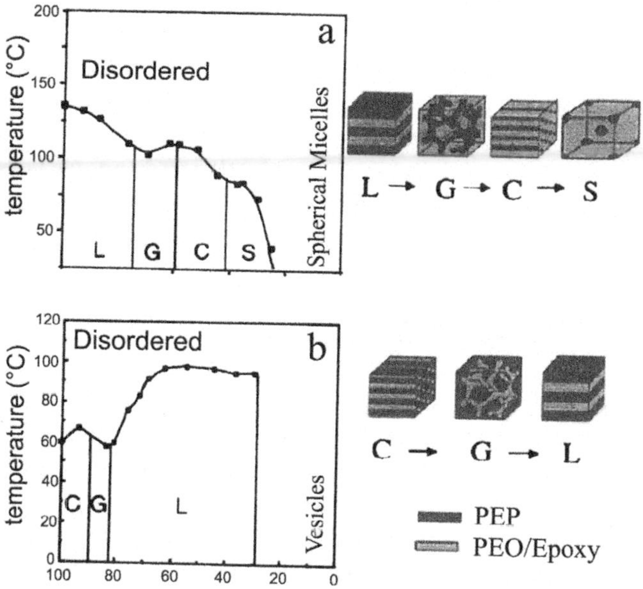

Fig. 66 Phase diagrams of **a** symmetric ($\phi_{PEO} = 0.51$, $M_n = 2700$, $M_w/M_n = 1.10$) and **b** asymmetric ($\phi_{PEO} = 0.32$, $M_n = 2100$, $M_w/M_n = 1.14$) PEO-*b*-PEP block copolymers blended with epoxy resin. Phase transitions which originate from swelling of PEO chains with epoxy and/or curing agent are drawn as *single lines*, without implication that there are no coexistence regions. From [197]. Copyright 2001 Wiley

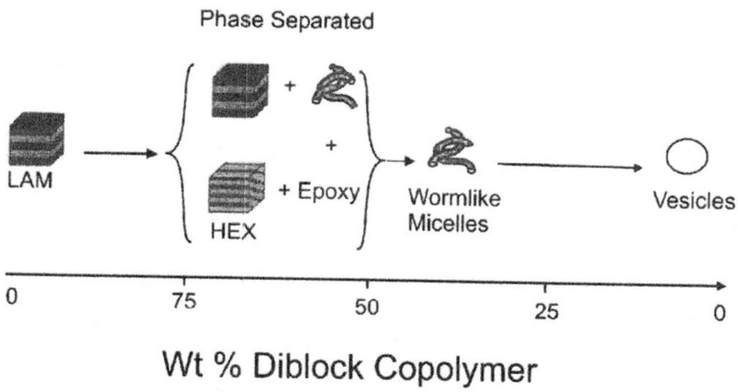

Fig. 67 Schematic of phase behaviour for blend of novolac epoxy resin with nearly symmetric poly(methyl acrylate-*co*-glycidylmethacrylate)-*b*-polyisoprene. Ordered *L* can be swollen with up to about 30% of resin before macroscopic phase separation occurs, producing heterogeneous morphologies containing various amounts of *L*, *C*, worm-like micelles and pristine epoxy. At lower concentrations, disordered worm-like micelles transform into vesicles in dilute limit. According to [201]. Copyright 2003 Wiley

in epoxy resins. During the curing process the block copolymer reacts with the resin and thus is prevented from macrophase separation. The block copolymer acts as a filler particle in the resin matrix.

Another way to blend reactively is to form the compatibilizing agent during the blending process. A very nice example was presented by Ruzette et al. [208], who blended poly(ethylene-*stat*-maleic anhydride) with a short- and long-chain polyamide (PA). During the melt processing, grafting of mainly short PA chains and the random copolymer occurred which self-assemble at the interface between the polyethylene and PA-rich domains. A cocontinuous morphology was obtained over a rather large region of composition and led to a very interesting thermoplastic material.

4.6 Influence of Nanoparticles on Phase Behaviour

The impact of nanoparticles on the phase behaviour of binary block copolymers has been treated by various authors. Balazs's group [209–213] employed a combination of SCFT for diblocks with a weighted-density approximation of a density functional theory for spherical particles to determine the distribution of particles within *L* morphologies. Due to self-assembly of the particles within the compatible domain, they predicted centre- and edge-filled *L* or *C* morphologies depending on the particle size and concentration. Their results indicated that these particles may localize in the minority phase and the fillers can form various structures ranging from nanosheets, nanowires and nanodots to continuous nanoarrays stemming from *PL* or *G* structures. If the particles are located in the majority phase of the system, a variety of different nanoporous materials are created.

Influence on the phase behaviour is observed if even small amounts (0.5 wt %) of hairy silica-like nanoparticles obtained by block-copolymer-directed sol-gel synthesis are added to a block copolymer [53, 214]. On the example of a PS-*b*-PI diblock copolymer a depression of the ODT of approx. 20 K was observed (Fig. 68), accompanied by a broadening of the transition. The largest depression was observed for rod-like nanoparticles, which is explained by the energy introduced by the defects [215].

Interestingly, the selective swelling of one phase of a block copolymer with a sol-gel precursor can lead to a variation in the microstructure. Finnefrock et al. [216] investigated the PI-*b*-PEO/aluminosilicate hybrid material morphology in an area of block copolymer phase space where a gyroid phase ($Ia\bar{3}d$ symmetry) was expected. However, TEM and 2*d*-SAXS data were not consistent with a gyroid but rather suggested a Plumber's Nightmare [217] morphology with $Im\bar{3}m$ symmetry.

The influence of magnetite nanoparticles on the lamellar dimensions of a symmetric deuterated polystyrene-*b*-poly(butyl methacrylate) film was investigated by neutron scattering. The magnetite particles exhibited an average diameter of 5 nm and were covered with PS hairs from sulfonated α-lithium PS to enable selective swelling of the *d*PS phase. Incorporation of 7 vol % magnetite particles resulted in a 12% increase of the *d*PS lamellae's

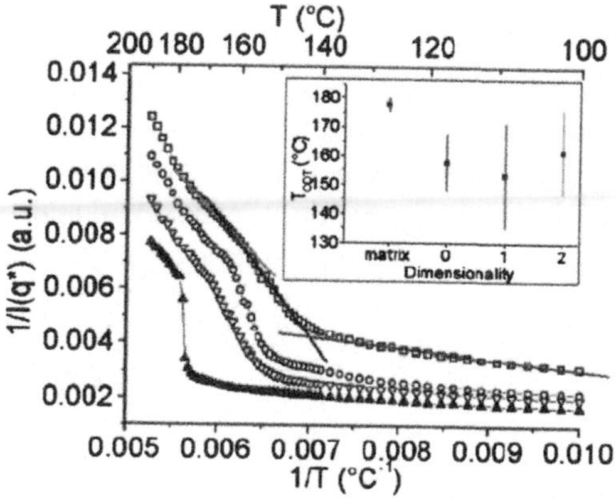

Fig. 68 Comparison of temperature-dependent intensity of first-order Bragg peak for bare matrix copolymer (▲) containing 0.5 wt % nanocomposites with plate-like (▽), spherical (○) and rod-like (□) geometry. Data are vertically shifted for clarity. *Inset*: dependence of ODT temperature on dimensionality of fillers (spherical 0, rod-like 1, plate-like 2). *Vertical bars*: width of phase transition region. Pure block copolymer is denoted "matrix". From [215]. Copyright 2003 American Chemical Society

thickness compared to the unfilled sample. Furthermore, the results indicate that still a higher amount of nanoparticles can be incorporated without destroying the phase separation [218].

5 Conclusion and Outlook

The structural characterization of block copolymers is an exciting and still emerging research area offering enormous scientific and technological promise. The choice of the appropriate chemistry, composition and architecture of the unlike blocks can lead to morphology control on the nanoscale. The generated structures can be further altered by selective treatment of one phase or by the use of physical forces, e.g. shear deformation or electrical fields. The future potential of the use of block copolymers for new material lies in the versatility of the polymer chemistry. Compared to linear AB diblock copolymers, the number of morphologies for linear ABC triblock terpolymers is more than doubled leading to nanostructures as complex as rings or even helices. More structural variations are accessible when the block polymers are blended or their architecture is changed from a linear array to a more complex one. The blending route sounds especially appealing as the properties of readily available and low-priced homopolymers may be tailored and refined by simple addition to a block copolymer.

Nanoengineering of such structures towards applications has been demonstrated in many areas. Some of them are presented in later articles of this volume such as patterning of surfaces (lithography) and template chemistry leading to 3D structure of materials which cannot self-assemble in such geometries by themselves. Nano-objects of predetermined size, shape and composition will find applications e.g. as (multi)functional additives in composite materials. Examples are their use in high-impact modification of polymers or compatibilization of polymer blends. The uses of block copolymers as membranes for separation technology and catalysis and, last but not least, for polymer-based photonic crystals are other areas in which technologies will develop. However, as a major challenge in some of these areas still remains the control of the orientation of the microstructures on larger length scales and suppression of morphological defects.

Acknowledgements The authors are indebted to Ms A. Boschetti de Fierro, Geesthacht for proofreading the manuscript and many helpful suggestions. Furthermore, they wish to thank P. Witt, Geesthacht, for his help in literature searches.

References

1. Riess G, Hurtrez G, Bahadur P (1985) Encycl Polym Sci Eng 2:324
2. Bates FS, Fredrickson GH (1990) Annu Rev Phys Chem 41:525
3. Hamley I (1998) The Physics of Block Copolymers. Oxford University Press, Oxford
4. Bates FS, Fredrickson GH (1999) Phys Today 52:32
5. Abetz V (2003) Block Copolymers, Ternary Triblocks. In: Kroschwitz JI (ed) Encyclopedia of Polymer Science and Technology, vol 1. Wiley, New York, p 482
6. Park C, Yoon J, Thomas E (2003) Polymer 44:6725
7. Ruzette A, Leibler L (2005) Nat Mater 4:19
8. Abetz V, Goldacker T (2000) Macromol Rapid Commun 21:16
9. Förster S, Khandpur AK, Zhao J, Bates FS, Hamley IW, Ryan AJ, Bras W (1994) Macromolecules 27:6922
10. Hajduk DA, Harper PE, Gruner SM, Honeker CC, Kim G, Thomas EL, Fetters LJ (1994) Macromolecules 27:4063
11. Khandpur AK, Foerster S, Bates FS, Hamley IW, Ryan AJ, Bras W, Almdal K, Mortensen K (1995) Macromolecules 28:8796
12. Hajduk DA, Takenouchi H, Hillmyer MA, Bates FS, Vigild ME, Almdal K (1997) Macromolecules 30:3788
13. Vigild ME, Almdal K, Mortensen K, Hamley IW, Fairclough JPA, Ryan AJ (1998) Macromolecules 31:5702
14. Ahn J-H, Zin W-C (2000) Macromolecules 33:641
15. Yang L, Gido SP, Mays JW, Pispas S, Hadjichristidis N (2001) Macromolecules 34:4235
16. Ho R-M, Chiang Y-W, Tsai C-C, Lin C-C, Ko B-T, Huang B-H (2004) J Am Chem Soc 126:2704
17. Leibler L (1980) Macromolecules 13:1602
18. Meier DJ (1969) J Polym Sci Part C Polym Symp 26PC:81
19. Semenov AN (1985) Zh Eksp Teor Fiz 88:1242

20. Matsen MW, Bates FS (1997) J Chem Phys 106:2436
21. Olvera de la Cruz M, Sanchez IC (1986) Macromolecules 19:2501
22. Floudas G, Hadjichristidis N, Iatrou H, Pakula T, Fischer EW (1994) Macromolecules 27:7735
23. Floudas G, Pispas S, Hadjichristidis N, Pakula T, Erukhimovich I (1996) Macromolecules 29:4142
24. Floudas G, Hadjichristidis N, Tselikas Y, Erukhimovich I (1997) Macromolecules 30:3090
25. Floudas G, Hadjichristidis N, Iatrou H, Avgeropoulos A, Pakula T (1998) Macromolecules 31:6943
26. Fredrickson GH, Helfand E (1987) J Chem Phys 87:697
27. Erukhimovich I, Dobrynin AV (1994) Macromol Symp 81:253
28. Vavasour JD, Whitmore MD (1993) Macromolecules 26:7070
29. Vavasour JD, Whitmore MD (1996) Macromolecules 29:5244
30. Matsen MW, Schick M (1994) Macromolecules 27:4014
31. Matsen MW, Bates FS (1997) J Polym Sci Part B Polym Phys 35:945
32. Matsen MW (2000) J Chem Phys 113:5539
33. Almdal K, Hillmyer MA, Bates FS (2002) Macromolecules 35:7685
34. Ko MJ, Kim SH, Jo WH (2001) Macromol Theory Simul 10:381
35. Ko MJ, Kim SH, Jo WH (2002) Fibers Polym 3:8
36. Ko MJ, Kim SH, Jo WH (2003) Fibers Polym 4:15
37. Tang P, Qiu F, Zhang H, Yang Y (2004) Phys Rev E Stat Nonlinear Soft Matter Phys 69:031803/1
38. Bohbot-Raviv Y, Wang Z (2000) Phys Rev Lett 85:3428
39. Davidock DA, Hillmyer MA, Lodge TP (2003) Macromolecules 36:4682
40. Davidock DA, Hillmyer MA, Lodge TP (2004) Macromolecules 37:397
41. Ren Y, Lodge TP, Hillmyer MA (2000) Macromolecules 33:866
42. Park S, Kwon K, Cho D, Lee B, Ree M, Chang T (2003) Macromolecules 36:4662
43. Spontak RJ, Smith SD (2001) J Polym Sci Part B Polym Phys 39:947
44. Wu L, Lodge TP, Bates FS (2004) Macromolecules 37:8184
45. Ganesan V, Fredrickson G (2001) J Rheol 45:161
46. Wang X, Dormidontova EE, Lodge TP (2002) Macromolecules 35:9687
47. Cavicchi KA, Lodge TP (2003) Macromolecules 36:7158
48. Cavicchi KA, Lodge TP (2004) Macromolecules 37:6004
49. Stepanek P, Nallet F, Diat O, Almdal K, Panine P (2002) Macromolecules 35:7287
50. Floudas G, Ulrich R, Wiesner U (1999) J Chem Phys 110:652
51. Floudas G, Ulrich R, Wiesner U, Chu B (2000) Europhys Lett 50:182
52. Floudas G, Vazaiou B, Schipper F, Ulrich R, Wiesner U, Iatrou H, Hadjichristidis N (2001) Macromolecules 34:2947
53. Simon PFW, Ulrich R, Spiess HW, Wiesner U (2001) Chem Mater 13:3464
54. Ryu DY, Jeong U, Kim JK, Russell TP (2002) Nat Mater 1:114
55. Ryu DY, Jeong U, Lee DH, Kim J, Youn HS, Kim JK (2003) Macromolecules 36:2894
56. Ryu DY, Lee DJ, Kim JK, Lavery KA, Russell TP, Han YS, Seong BS, Lee CH, Thiyagarajan P (2003) Phys Rev Lett 90:235501
57. Ryu DY, Lee DH, Jeong U, Yun SH, Park S, Kwon K, Sohn BH, Chang T, Kim JK (2004) Macromolecules 37:3717
58. Ryu DY, Park MS, Chae SH, Jang J, Kim JK, Russell TP (2002) Macromolecules 35:8676
59. Ryu DY, Lee DH, Jang J, Kim JK, Lavery KA, Russell TP (2004) Macromolecules 37:5851

60. Kim JK, Jang J, Lee DH, Ryu DY (2004) Macromolecules 37:8599
61. Ryan AJ, Mai S-M, Fairclough JPA, Hamley IW, Booth C (2001) Phys Chem Chem Phys 3:2961
62. Matsen MW (1995) J Chem Phys 102:3884
63. Schmidt SC, Hillmyer MA (2002) J Polym Sci Part B Polym Phys 40:2364
64. Ludwigs S, Boeker A, Abetz V, Muller AHE, Krausch G (2003) Polymer 44:6815
65. Ludwigs S, Boeker A, Abetz V, Mueller AHE, Krausch G (2002) Polym Prepr 43:377
66. Elbs H, Abetz V, Hadziioannou G, Drummer C, Krausch G (2001) Macromolecules 34:7917
67. Fukunaga K, Hashimoto T, Elbs H, Krausch G (2002) Macromolecules 35:4406
68. Fukunaga K, Hashimoto T, Elbs H, Krausch G (2003) Macromolecules 36:2852
69. Ludwigs S, Boeker A, Voronov A, Rehse N, Magerle R, Krausch G (2003) Nat Mater 2:744
70. Elbs H, Drummer C, Abetz V, Krausch G (2002) Macromolecules 35:5570
71. Elbs H, Fukunaga K, Stadler R, Sauer G, Magerle R, Krausch G (1999) Macromolecules 32:1204
72. Rehse N, Knoll A, Magerle R, Krausch G (2003) Macromolecules 36:3261
73. Neumann C, Loveday DR, Abetz V, Stadler R (1998) Macromolecules 31:2493
74. Hashimoto T, Yamauchi K, Yamaguchi D, Hasegawa H (2003) Macromol Symp 201:65
75. Epps TH, Cochran EW, Bailey TS, Waletzko RS, Hardy CM, Bates FS (2004) Macromolecules 37:8325
76. Epps TH, Cochran EW, Hardy CM, Bailey TS, Waletzko RS, Bates FS (2004) Macromolecules 37:7085
77. Bailey TS, Hardy CM, Epps TH, Bates FS (2002) Macromolecules 35:7007
78. Shefelbine TA (1999) J Am Chem Soc 121:8457
79. Seki M, Suzuki J, Matsushita Y (2000) J Appl Crystallogr 33:285
80. Suzuki J, Seki M, Matsushita Y (2000) J Chem Phys 112:4862
81. Mogi Y, Mori K, Matsushita Y, Noda I (1992) Macromolecules 25:5412
82. Cochran E, Bates F (2004) Phys Rev Lett 93
83. Matsen MW (1998) J Chem Phys 108:785
84. Gido SP, Schwark DW, Thomas EL, do Carmo Goncalves M (1993) Macromolecules 26:2636
85. Mogi Y, Mori K, Kotsuji H, Matsushita Y, Noda I, Han CC (1993) Macromolecules 26:5169
86. Hückstadt H, Göpfert A, Abetz V (2000) Polymer 41:9089
87. Bailey TS, Pham HD, Bates FS (2001) Macromolecules 34:6994
88. Hardy CM, Bates FS, Kim M-H, Wignall GD (2002) Macromolecules 35:3189
89. Takano A, Soga K, Suzuki J, Matsushita Y (2003) Macromolecules 36:9288
90. Jaffer KM, Wickham RA, Shi AC (2004) Macromolecules 37:7042
91. Takahashi K, Hasegawa H, Hashimoto T, Bellas V, Iatrou H, Hadjichristidis N (2002) Macromolecules 35:4859
92. Takano A, Soga K, Asari T, Suzuki J, Arai S, Saka H, Matsushita Y (2003) Macromolecules 36:8216
93. Mackay ME, Hong Y, Jeong M, Tande BM, Wagner NJ, Hong S, Gido SP, Vestberg R, Hawker CJ (2002) Macromolecules 35:8391
94. Gido SP, Wang ZG (1997) Macromolecules 30:6771
95. Marko JF (1993) Macromolecules 26:1442
96. Lecommandoux S, Borsali R, Schappacher M, Deffieux A, Narayanan T, Rochas C (2004) Macromolecules 37:1843

97. Takano A, Kadoi O, Hirahara K, Kawahara S, Isono Y, Suzuki J, Matsushita Y (2003) Macromolecules 36:3045
98. Mai S-M, Fairclough JPA, Terrill NJ, Turner SC, Hamley IW, Matsen MW, Ryan AJ, Booth C (1998) Macromolecules 31:8110
99. Jo WH, Jang SS (1999) J Chem Phys 111:1712
100. Zhu Y, Gido SP, Iatrou H, Hadjichristidis N, Mays JW (2003) Macromolecules 36:148
101. Adhikari R, Michler GH, Lebek W, Goerlitz S, Weidisch R, Knoll K (2002) J Appl Polym Sci 85:701
102. Adhikari R, Michler GH, Huy TA, Ivan'kova E, Godehardt R, Lebek W, Knoll K (2003) Macromol Chem Phys 204:488
103. Adhikari R, Michler GH, Godehardt R, Ivan'kova EM (2003) Polymer 44:8041
104. Winey KI, Thomas EL, Fetters LJ (1991) J Chem Phys 95:9367
105. Winey KI, Thomas EL, Fetters LJ (1992) Macromolecules 25:2645
106. Koizumi S, Hasegawa H, Hashimoto T (1992) Makromol Chem Macromol Symp 62:75
107. Koizumi S, Hasegawa H, Hashimoto T (1994) Macromolecules 27:6532
108. Adhikari R, Michler GH, Goerlitz S, Knoll K (2004) J Appl Polym Sci 92:1208
109. Adhikari R, Michler GH, Henning S, Godehardt R, Huy TA, Goerlitz S, Knoll K (2004) J Appl Polym Sci 92:1219
110. Adhikari R, Michler GH, Knoll K (2004) Polymer 45:241
111. Milner ST (1994) Macromolecules 27:2333
112. Grason GM, Kamien RD (2004) Macromolecules 37:7371
113. Mavroudis A, Avgeropoulos A, Hadjichristidis N, Thomas EL, Lohse DJ (2003) Chem Mater 15:1976
114. Zhongde X, Mays J, Xuexin C, Hadjichristidis N, Schilling FC, Bair HE, Pearson DS, Fetters LJ (1985) Macromolecules 18:2560
115. Hückstädt H, Goldacker T, Göpfert A, Abetz V (2000) Macromolecules 33:3757
116. Birshtein TM, Polotsky AA, Abetz V (2004) Macromol Theory Simul 13:512
117. Hückstädt H, Göpfert A, Abetz V (2000) Macromol Chem Phys 201:296
118. Torikai N, Matsushita Y, Langridge S, Bucknall D, Penfold J, Takeda M (2000) Phys B (Amsterdam) 283:12
119. Grayer V, Dormidontova EE, Hadziioannou G, Tsitsilianis C (2000) Macromolecules 33:6330
120. Buzza DMA, Fzea AH, Allgaier JB, Young RN, Hawkins RJ, Hamley IW, McLeish TCB, Lodge TP (2000) Macromolecules 33:8399
121. Zhu Y, Gido SP, Moshakou M, Iatrou H, Hadjichristidis N, Park S, Chang T (2003) Macromolecules 36:5719
122. Avgeropoulos A, Dair BJ, Thomas EL, Hadjichristidis N (2002) Polymer 43:3257
123. Fredrickson GH (1991) Macromolecules 24:3456
124. Olvera de la Cruz M, Mayes AM, Swift BW (1992) Macromolecules 25:944
125. Mays JW, Uhrig D, Gido S, Zhu YQ, Weidisch R, Iatrou H, Hadjichristidis N, Hong K, Beyer F, Lach R, Buschnakowski M (2004) Macromol Symp 215:111
126. Huo Y, Zhang H, Yang Y (2003) Macromolecules 36:5383
127. Tanaka H (1997) Phys Rev E 56:4451
128. Cavicchi KA, Lodge TP (2003) J Polym Sci Part B Polym Phys 41:715
129. Dormidontova EE, Lodge TP (2001) Macromolecules 34:9143
130. Sakamoto N, Hashimoto T, Han CD, Kim D, Vaidya NY (1997) Macromolecules 30:1621

131. Kim JK, Lee HH, Sakurai S, Aida S, Masamoto J, Nomura S, Kitagawa Y, Suda Y (1999) Macromolecules 32:6707
132. Han CD, Vaidya NY, Kim D, Shin G, Yamaguchi D, Hashimoto T (2000) Macromolecules 33:3767
133. Choi S, Vaidya NY, Han CD, Sota N, Hashimoto T (2003) Macromolecules 36:7707
134. Riise BL, Fredrickson GH, Larson RG, Pearson DS (1995) Macromolecules 28:7653
135. Sota N, Sakamoto N, Saijo K, Hashimoto T (2003) Macromolecules 36:4534
136. Kimishima K, Koga T, Hashimoto T (2000) Macromolecules 33:968
137. Wang C-Y, Lodge TP (2002) Macromolecules 35:6997
138. Matsen MW, Bates FS (1996) Macromolecules 29:1091
139. Lee HH, Jeong W-Y, Kim JK, Ihn KJ, Kornfield JA, Wang Z-G, Qi S (2002) Macromolecules 35:785
140. Qi SY, Wang ZG (1998) Polymer 39:4639
141. Ryu CY, Vigild ME, Lodge TP (1998) Phys Rev Lett 81:5354
142. Ryu CY, Lodge TP (1999) Macromolecules 32:7190
143. Krishnamoorti R, Modi MA, Tse MF, Wang H-C (2000) Macromolecules 33:3810
144. Krishnamoorti R, Silva AS, Modi MA (2000) Macromolecules 33:3803
145. Koppi KA, Tirrell M, Bates FS, Almdal K, Mortensen K (1994) J Rheol 38:999
146. Ryu CY, Lee MS, Haiduk DA, Lodge TP (1997) J Polym Sci Part B Polym Phys 35:2811
147. Kim JK, Lee HH, Gu Q-J, Chang T, Jeong JH (1998) Macromolecules 31:4045
148. Kim JK, Lee HH, Ree M, Lee K-B, Park Y (1998) Macromol Chem Phys 199:641
149. Wang C-Y, Lodge TP (2002) Macromol Rapid Commun 23:49
150. Schulz MF, Bates FS, Almdal K, Mortensen K (1994) Phys Rev Lett 73:86
151. Hajduk DA, Ho R-M, Hillmyer MA, Bates FS, Almdal K (1998) J Phys Chem B 102:1356
152. Ahn J-H, Zin W-C (2003) Macromol Res 11:152
153. Ahn J-H, Zin W-C (2002) Macromolecules 35:10238
154. Chastek TQ, Lodge TP (2003) Macromolecules 36:7672
155. Goveas JL, Milner ST, Russel WB (1997) Mater Res Soc Symp Proc 463:167
156. Goveas JL, Milner ST (1997) Macromolecules 30:2605
157. Bang J, Lodge TP (2003) J Phys Chem B 107:12071
158. Sakurai S, Kota T, Isobe D, Okamoto S, Sakurai K, Ono T, Imaizumi K, Nomura S (2004) J Macromol Sci Phys B43:1
159. Suzuki J, Furuya M, Iinuma M, Takano A, Matsushita Y (2002) J Polym Sci Part B Polym Phys 40:1135
160. Matsushita Y, Noro A, Iinuma M, Suzuki J, Ohtani H, Takano A (2003) Macromolecules 36:8074
161. Lodge TP, Hanley KJ, Pudil B, Alahapperuma V (2003) Macromolecules 36:816
162. Hanley KJ, Lodge TP, Huang C-I (2000) Macromolecules 33:5918
163. Lodge TP, Pudil B, Hanley KJ (2002) Macromolecules 35:4707
164. Shusharina NP, Alexandridis P, Linse P, Balijepalli S, Gruenbauer HJM (2003) Eur Phys J E 10:45
165. Abetz V, Markgraf K, Rebizant V (2002) Macromol Symp 177:139
166. Ott H, Abetz V, Altstaedt V (2001) Macromolecules 34:2121
167. Drummy LF, Voigt-Martin I, Martin DC (2001) Macromolecules 34:7416
168. Breiner U, Krappe U, Stadler R (1996) Macromol Rapid Commun 17:567

169. Breiner U, Krappe U, Thomas EL, Stadler R (1998) Macromolecules 31:135
170. Epps TH, III, Bailey TS, Pham HD, Bates FS (2002) Chem Mater 14:1706
171. Epps TH, III, Bailey TS, Waletzko R, Bates FS (2003) Macromolecules 36:2873
172. Whitmore MD, Noolandi J (1985) Macromolecules 18:2486
173. Matsen MW (1995) Macromolecules 28:5765
174. Bodycomb J, Yamaguchi D, Hashimoto T (2000) Macromolecules 33:5187
175. Mykhaylyk TA, Mykhaylyk OO, Collins S, Hamley IW (2004) Macromolecules 37:3369
176. Schulz MF, Khandpur AK, Bates FS, Almdal K, Mortensen K, Hajduk DA, Gruner SM (1996) Macromolecules 29:2857
177. Sugiyama M, Shefelbine TA, Vigild ME, Bates FS (2001) J Phys Chem B 105:12448
178. Yamaguchi D, Shiratake S, Hashimoto T (2000) Macromolecules 33:8258
179. Yamaguchi D, Hasegawa H, Hashimoto T (2001) Macromolecules 34:6506
180. Yamaguchi D, Hashimoto T (2001) Macromolecules 34:6495
181. Yamaguchi D, Takenaka M, Hasegawa H, Hashimoto T (2001) Macromolecules 34:1707
182. Almdal K, Rosedale JH, Bates FS (1990) Macromolecules 23:4336
183. Floudas G, Vlassopoulos D, Pitsikalis M, Hadjichristidis N, Stamm M (1996) J Chem Phys 104:2083
184. Matsen MW (1995) J Chem Phys 103:3268
185. Matsen MW, Bates FS (1995) Macromolecules 28:7298
186. Court F, Hashimoto T (2001) Macromolecules 34:2536
187. Kane L, Norman DA, White SA, Matsen MW, Satkowski MM, Smith SD, Spontak RJ (2001) Macromol Rapid Commun 22:281
188. Goldacker T, Abetz V, Stadler R, Erukhimovich I, Leibler L (1999) Nature 398:137
189. Goldacker T, Abetz V, Stadler R (2000) Macromol Symp 149:93
190. Abetz V, Jiang S, Gopfert A (2004) E-Polymers 40
191. Wickham RA, Shi AC (2001) Macromolecules 34:6487
192. Jiang S, Goepfert A, Abetz V (2003) Macromol Rapid Commun 24:932
193. Abetz V, Jiang S (2004) E-Polymers 54
194. Jiang S, Goepfert A, Abetz V (2003) Macromolecules 36:6171
195. Auschra C, Stadler R (1993) Macromolecules 26:6364
196. Kirschnick T, Gottschalk A, Ott H, Abetz V, Puskas J, Altstadt V (2004) Polymer 45:5653
197. Dean JM, Lipic PM, Grubbs RB, Cook RF, Bates FS (2001) J Polym Sci Part B Polym Phys 39:2996
198. Guo QP, Peng XS, Wang ZJ (1991) Polymer 32:53
199. Sixun Z, Naibin Z, Xiaolie L, Dezhu M (1995) Polymer 36:3609
200. Lipic PM, Bates FS, Hillmyer MA (1998) J Am Chem Soc 120:8963
201. Guo Q, Dean JM, Grubbs RB, Bates FS (2003) J Polym Sci Part B Polym Phys 41:1994
202. Kinning DJ, Winey KI, Thomas EL (1988) Macromolecules 21:3502
203. Zheng Y, Lin Z, Zakin JL, Talmon Y, Davis HT, Scriven LE (2000) J Phys Chem B 104:5263
204. Utiyama H, Takenaka K, Mizumori M, Fukuda M, Tsunashi Y, Kurata M (1974) Macromolecules 7:515
205. Antonietti M, Heinz S, Schmidt M, Rosenauer C (1994) Macromolecules 27:3276
206. Rebizant V, Abetz V, Tournilhac F, Court F, Leibler L (2003) Macromolecules 36:9889

207. Rebizant V, Venet A, Tournilhac F, Girard-Reydet E, Navarro C, Pascault J, Leibler L (2004) Macromolecules 37:8017
208. Ruzette A (2002) Nat Mater 1:85
209. Thompson RB, Ginzburg VV, Matsen MW, Balazs AC (2001) Science 292:2469
210. Thompson RB, Ginzburg VV, Matsen MW, Balazs AC (2002) Macromolecules 35:1060
211. Lee J-Y, Thompson RB, Jasnow D, Balazs AC (2002) Phys Rev Lett 89:155503/1
212. Lee JY, Thompson RB, Jasnow D, Balazs AC (2003) Faraday Discuss 123:121
213. Balazs AC (2003) Curr Opin Solid State Mater Sci 7:27
214. Ulrich R, Du Chesne A, Templin M, Wiesner U (1999) Adv Mater 11:141
215. Jain A, Gutmann JS, Garcia CBW, Zhang Y, Tate MW, Gruner SM, Wiesner U (2002) Macromolecules 35:4862
216. Finnefrock AC, Ulrich R, Du Chesne A, Honeker CC, Schumacher K, Gruner SM, Wiesner U (2001) Angew Chem Int Ed Engl 40:1207
217. Huse DA, Leibler S (1988) J Phys 49:605
218. Lauter-Pasyuk V, Lauter HJ, Gordeev GP, Mueller-Buschbaum P, Toperverg BP, Jernenkov M, Petry W (2003) Langmuir 19:7783
219. Yang L, Hong S, Gido SP, Velis G, Hadjichristidis N (2001) Macromolecules 34:9069

Author Index Volumes 101–189

Author Index Volumes 1–100 see Volume 100

Subject Index